ENGINEERING

THE

ANCIENT
WORLD

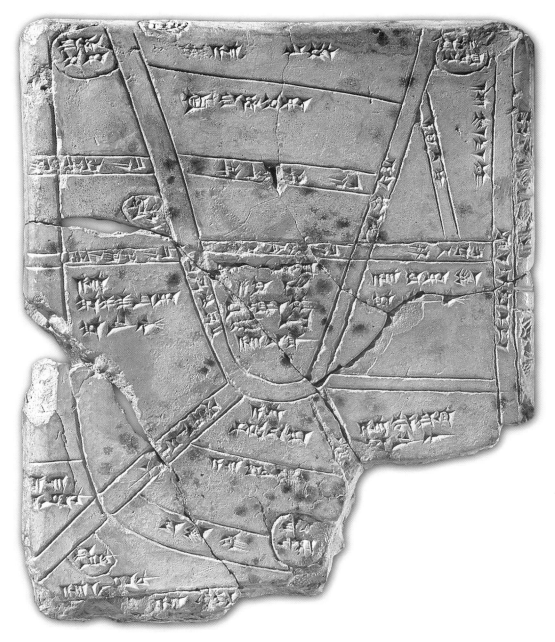

Baked clay tablet found at Nippur in Mesopotamia, dating from about 1300 BC.

ENGINEERING

THE

ANCIENT

WORLD

DICK PARRY

SUTTON PUBLISHING

First published in the United Kingdom in 2005 by
Sutton Publishing Limited · Phoenix Mill
Thrupp · Stroud · Gloucestershire · GL5 2BU

British Library Cataloguing in Publication Data
A catalogue record for this book is available from the British Library.

ISBN 0-7509-3833-1

Typeset in 10/13 pt Sabon.
Typesetting and origination by
Sutton Publishing Limited.
Printed in Great Britain by
Printed and bound in England by
J.H. Haynes & Co. Ltd, Sparkford.

CONTENTS

ACKNOWLEDGEMENTS

The full-scale rolling tests to explore this possible method for transporting and raising Pyramid stones were financed and carried out in Japan by the Obayashi Corporation in 1996, while those relating to the Stonehenge bluestones were carried out in 1998 on the mountainside of Carne Meini, immediately below the bluestone outcrop, as part of the TV programme *Stonehenge: Secrets of the Stones* produced by Yorkshire Television. The many extracts from Herodotus, *The Histories*, are reproduced with the kind permission of Penguin Books Limited, and I am grateful, too, to Cambridge University Press for permission to include extracts from *Science and Civilisation in China* by Joseph Needham, Lu Gwei-Djen and Ling Wang. I am much indebted to my good friend Demetrios Coumoulos for assisting me in obtaining information on the Eupalinos tunnel. My thanks also to Christopher Feeney, Hilary Walford, Jane Entrican, Victoria Carvey and other staff members at Sutton Publishing who have contributed to making the book a reality.

The publishers are also grateful to the following:

AA&A Collection (pp. 117, 172, 182, 209)
AGE Fotostock (p. 75)
Ashmolean Museum, Oxford (pp. 13 (top), 181)
Cambridge University (pp. 104, 140)
Department of Environment, Heritage and Local Government, Dublin (p. 197)
English Heritage (pp. 146, 157 (bottom))
Mary Evans Picture Library (p. 68)
The Fotomas Index (p. 93 (bottom))
Foreign Languages Press, China (pp. 122, 123)
Friedrich-Schiller-Universtät, Jena (p. 40)
Sonia Halliday Photographs (pp. 14, 23 (top), 23 (bottom), 33, 55, 87, 88, 112, 116, 189, 191, 192, 193, 208, 210, 213, 216, 223)
Robert Harding Picture Library (pp. 31, 101, 108, 124, 125, 126, 179, 195, 196)
Jason Hawkes Aerial Photography Library (pp. 157 (top), 199)
Hellenic Republic, Ministry of Culture (p. 35)
National Trust (Alexander Keiller Museum) (p. 144)
© Caroline Malone, reproduced by permission of Batsford, an imprint of Chrysalis Book Group Plc (p. 141)

Adrian Parry (pp. 119, 120)

Topfoto/Woodmansterne (p. 74)

University of Pennsylvania Museum (pp. ii, 20)

Wonders of the Past (Hammerton) (pp. 59, 60 (bottom), 76, 107, 155, 156, 165 (top), 166, 174, 185)

All other images provided by the author.

PREFACE

This book owes its existence largely to a man who lived nearly 2,500 years ago. When gathering material for my book *Engineering the Pyramids* I found that the best description of the construction of the Great Pyramid, or anyway the one that convinced me the most, was that by the Greek historian Herodotus, writing in the fifth century BC. But, perhaps more importantly, I came to realise that our understanding of a number of the engineering achievements of our ancient forebears owes much to the writings of Herodotus. Most of those writing about Herodotus, whether laudatory in their comments or otherwise, have been classicists, and, not surprisingly, the engineering achievements of the time have not been their primary consideration. We are fortunate that Herodotus was a keen observer, had a curious mind, and had the ability to glean information from the priests and other informed people he met or sought out during his extensive travels. Cicero saw him as the 'Father of History' and, as he might well be the most quoted (and misquoted) author in any field, this seems to be a reasonable claim; but there have been, and still are, those who view him in a different light. He was certainly a teller of tall tales, but he invariably makes it clear that he is simply repeating something told to him: a modern historian would be expected to judge such bits of information critically and weed them out, or suffer the opprobrium of his peers; but Herodotus, free of precedent and peers, leaves the reader to make what he or she will of them.

The world Herodotus knew, or knew about, comprised the countries fringing the Mediterranean, and extending eastward at least as far as the Persian Gulf. He also had some acquaintance with areas fringing the Black Sea. His knowledge of the geography and history of these areas came from his own observations and discussions with local people, other travellers and learned persons, such as priests, during his travels and to a limited extent from existing texts, such as the writings of Hecateus. It is clear from his many references to ancient works that he recognised that the world he knew owed much to the engineers who had devised the means to cross rivers, traverse the land, irrigate the crops, supply water for human construction, and build the great temples and tombs. It was these engineers, and not the great kings or their armies, who established the very foundations of civilisation. An appreciation of the work of these ancient engineers must not be limited to the works described or mentioned by Herodotus; but it is he who provides us with an excellent starting point from which we can look back to the great accomplishments preceding his time (many of which still existed in, or influenced, the world he knew), and forward to the great accomplishments that came after his time, which, in some instances, owed much to the knowledge accumulated before and during his time.

During his lifetime much of the world known to Herodotus, outside Greece itself, was under Persian control. But it was a world that over the previous three millennia had experienced a kaleidoscope of changing civilisations, each benefiting from the knowledge and technological expertise of its predecessors and sometimes its neighbours; each, in turn, advertently or inadvertently passing knowledge to later successors. Born in Halicarnassus around 480 BC, at a time when it was under Persian control, Herodotus moved to the island of Samos as a young man, perhaps frustrated by the heavy hand of authority in his home town. His interests during his lifetime ranged far and wide, and, in addition to his account of the construction of the Great Pyramid, he has given us descriptions of engineering works as diverse as the ancient Suez Canal, Lake Moeris and its water supply from the Nile, the tunnel of Eupalinos on Samos, the pontoon bridges of Darius and Xerxes constructed to cross the Bosphorus and Hellespont respectively, diversion of the Euphrates at Babylon, the great walls, moat and ziggurat of Babylon, and the great sepulchral mound of Alyattes. One of his descriptions, of a labyrinth close to Lake Moeris, has been the subject of speculation and discussion in modern times. Despite his claim that it surpassed the pyramids themselves, there is no evidence that such a remarkable building existed, although the site can be identified and some sort of structure certainly occupied it. Diodorus also gives a description of the labyrinth, which he believed to be a common tomb for twelve leaders who ruled simultaneously as kings in Egypt.

It was several hundred years after Herodotus that other writers such as Diodorus Siculus (*c.* 80 BC–20 BC) and Strabo (*c.* 64 BC–AD 25) gave similar prominence to engineering achievements in their writings. It was also in the first century BC that Vitruvius produced his *Ten Books of Architecture*, essential reading not only for Roman engineers and architects, but also for Renaissance luminaries such as Alberti, Bramante, Michelangelo and Palladio. Vitruvius lists the three essentials of good building as durability, convenience (i.e. function) and beauty, but also stresses the importance of economy, which he denotes as the proper management of materials and of site, as well as a thrifty balancing of cost and common sense in the construction of works. In the first century AD, Frontinus, Roman Consul, onetime Governor of Britain and later to occupy the high office of Manager of Aqueducts at Rome, produced his treatise on the *Aqueducts of Rome*.

The remarkable civil engineering achievements of the ancient world were not, of course, confined to the world known to Herodotus. The great megalithic structures of Malta and Western Europe pre-dated the Greek historian by 2,000 years and more. Great irrigation schemes and other works were instigated in China and Sri Lanka. Herodotus knew of a country far to the east inhabited by strange animals and humans with strange habits, but certainly did not know of the advanced Harappan civilisation situated along the Indus River, which pre-dated him by 1,500 years.

In ancient China two of the greatest achievements a young man could aspire to were to be able to write, preferably poetry, and to be able to construct a canal. As a result this produced some outstanding writers and some of the most outstanding hydraulic engineers in the ancient world. Unfortunately, literary and technical abilities seldom occurred in combination in single individuals, as noted by Needham:

What is true of living humanists in the West is also true of some of the Chinese scholars of long ago whose writings are often our only means of access to the techniques of past ages. The artisans and technicians knew very well what they were doing, but they were liable to be illiterate, or at least inarticulate. The bureaucratic scholars, on the other hand, were highly articulate but too often despised the rude mechanicals whose activities, for one reason or another, they wrote about from time to time. Thus even the authors whose words are now so precious were often more concerned with their literary style than with the details of the machines and processes that they mentioned. This superior attitude was also not unknown amongst the artists, back-room experts (like the mathematicians) of the officials' yamens [offices], so that often they were more interested in making a charming picture than in showing the precise details of machinery when they were asked to limn it, and now sometimes it is only by comparing one drawing with another that we can reach certainty about the technical content. At the same time there were many great scholar-officials throughout Chinese history from Chang Heng in the Han to Shen Kua in the Sung and Tai Chen in the Ch'ing who combined a perfect expertise in classical literature with complete mastery of the sciences of their day and the applications of these in artisanal practice.

For all these reasons our knowledge of the development of technology is still in a lamentably backward state, vital though it is for economic history, that broad meadow of flourishing speculation.

One might be excused for wondering if the man who wrote the above, pessimistic, account was the same one who produced the great tomes constituting *Science and Civilisation in China*, arguably the most authoritative account ever written in the fields of engineering and science history. Fortunately, many of the civil engineering achievements of the ancient Chinese remain substantially intact and in some cases still in use, such as the Anchi Bridge (unmatched in Europe for 1,000 years), the Kuanhsein irrigation project, the Grand Canal and substantial lengths of the Great Wall.

Structures which have remained intact, or substantially so, since ancient times do not necessarily easily yield up their secrets, such as their purposes or methods of construction, in the absence of contemporary written or pictorial evidence. There are no better examples of this than the Egyptian pyramids, which raise innumerable questions. Their purpose may seem obvious, but in how many were the remains of pharaohs actually entombed? Why was the pyramid shape adopted? What was their astronomical significance? Why do they differ so much in detail, for example in the locations of the tomb chambers? How were they built? The last question has attracted many suggested solutions, ranging from the credible through the illogical to the impossible, often disregarding simple principles of mechanics. Similar questions can be posed with respect to the great Neolithic and early Bronze Age stone structures of Western Europe, constructed at much the same time as the pyramids. Sufficient of Stonehenge remains to give a good indication of its completed form, if, in fact, it ever was completed; but what was the significance of its alignment with the solstices? How was this alignment exploited? Why was this site chosen, so far from the sources of the megaliths used in its construction? How were the megaliths, up to 40 tons in weight, transported to the site, then raised to the vertical and topped with lintels?

But perhaps the greatest Neolithic puzzle of them all is the purpose behind the construction of Silbury Hill, an explanation for which is attempted in this book.

In considering the title of this book, the reader might well ask what period comprises 'ancient'. In fact it cannot be simply a chronological delineation: social structures and technological development are other legitimate criteria. The civilisations of South and Central America, and the early Indian settlements in the USA such as at Cahokia, had more in common with early Mediterranean civilisations than they did with their contemporaries in Europe and consequently find a place in this book. Major engineering works and buildings built in the name of, or influenced by, the Christian and Islamic faiths, dating back to the conversion of Constantine in the former case and from the seventh century in the latter case, have not been included in this book and are deserving of their own volume.

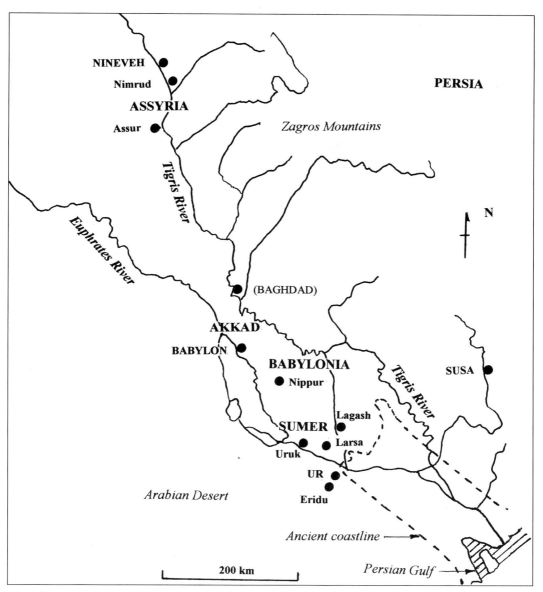

Plan of ancient Mesopotamia showing some of the principal cities.

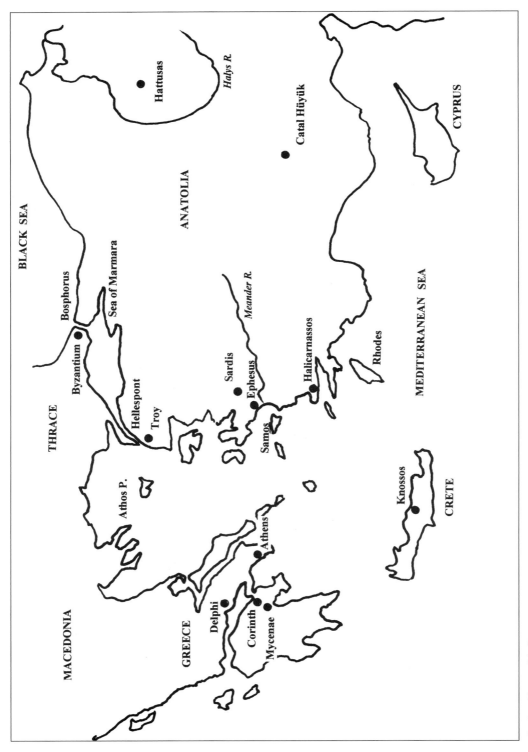

Ancient Eastern Mediterranean

1

CONTROLLING AND CONSUMING THE WATERS

Neolithic hunter-gatherers living on the Nile river plains and adjacent areas evolved around 6000 BC into farming communities able to exploit the rich agricultural and pastoral land bordering the river, which was replenished and revitalised each year by the annual inundation. Two distinct groupings emerged: a northern group centred around the modern Cairo–Fayum area (Lower Egypt) and a southern group (Upper Egypt). The predynastic period saw some merging of the two cultures, but also military conflicts, which culminated, around 3100 BC, in a victory by the king of Upper Egypt, Menes (also known as Narmer), who brought about the unification of the two states under a single king or pharaoh. However, clear distinctions between the two groups remained throughout pharaonic times and were recognised in the form of separate administrations and the pharaoh wearing the double crown. This comprised the pharaoh wearing both the high conical hat, or white crown of Upper Egypt, and the flat-topped cap with a tall projection at the back and a long feather curling forward – the red crown of Lower Egypt. The victory of Menes/Narmer and the subsequent unification is depicted on a famous schist palette that shows the king, wearing the white crown, smashing the skull of an adversary. On the other side, wearing the red crown, he is shown in regal marching pose preceded by the standard-bearers of the conquering nomes.

Around 3100 BC Menes established his capital at Memphis, 24km south of modern Cairo, having had the course of the Nile diverted to create a site suitable for a city replete with gardens, temples and palaces. The site was close to where the elongated narrow Nile valley of the Upper, or southern, largely arid region meets the fan-shaped Lower, or northern, productive marshy Delta, through which the river divides into several branches.

In order to ensure the continued unification of the two very different regions of the country, Menes put in hand major construction works, which would have occupied a workforce of many thousands. This was a political decision cementing the concept of unification. Herodotus describes the work as told to him:

> The priests told me that it was Min [Menes], the first king of Egypt, who raised the dam which created Memphis. The river used to flow along the base of the sandy hills on the Libyan border, and this monarch, by damning it up at the bend about a hundred

furlongs south of Memphis, drained the original channel and diverted it to a new one half-way between the two lines of hills. To this day the elbow which the Nile forms here, where it is forced into its new channel, is most carefully watched by the Persians, who strengthen the dam every year; for should the river burst it, Memphis might be completely overwhelmed. On the land which had been drained by the diversion of the river, King Min built the city which is now called Memphis – it lies in the narrow part of Egypt – and afterwards on the north and west sides of the town excavated a lake, communicating with the river, which itself protects it on the east. In addition to this the priests told me that he built there the large and very remarkable temple of Hephaestus.

It is no coincidence that the early civilisations developed in Mesopotamia and Egypt. The interrelationships and interdependencies between humans, which are the basis of civilisation and urban life, depend, more than any other factor, on the ready availability of water and the ability to control and exploit it. The Euphrates and Tigris (and their tributaries and lesser rivers in Mesopotamia), and the Nile, provided, most of the time at least, a reliable abundance of this commodity: it was simply a matter of controlling and exploiting this largesse.

The rivers that gave could also take away. At its highest levels in June and July, after heavy rains or melting snow at its source in the Turkish mountains, and possibly boosted by high tides in the Persian Gulf, the water level in the Euphrates exceeded by several metres the level of the surrounding land. Any breaching of the banks could lead to widespread and devastating floods, covering the land for months. But much more insidious was the river behaviour that led to the ultimate demise of many of the great city states of the Mesopotamian plains. A great river winding its way through plains such as these removes the highly erodable alluvial deposits on the outside of its bends and deposits them inside bends downstream, or along stretches where the water velocity drops markedly, thus building up its bed. Over a period of time the process of erosion and deposition can lead, inevitably, to changes amounting to tens of kilometres more in the course of the river. Rampaging floodwaters can also cut new channels. Settlements deprived of the river, and depending on it for their very existence, cannot survive. Woolley's excavations showed the Euphrates to have 'washed the walls of Ur on the west'. From the river, canals led into the city conveying water-borne traffic, and into the fields, spreading far across the plains, for irrigation. Today the river runs 16km to the east of the ruins and the great plain is a barren desert.

In ancient societies, women, with only a few exceptions such as Boudicca in Britain or Hatshepsut in Egypt, had little influence on administration or religion, or in the conduct of wars. If some of the ancient writers are to be believed, Semiramis of Assyria was also such an exception, although the truth seems to be that she was a semi-fictitious figure based on Sammuramat, an Assyrian queen who acted as regent for a few years until her son Adad-nirari III came of age. She may well have numbered some major achievements during her short regency, but certainly not those attributed to her by Diodorus, or, perhaps more specifically, by Ctesias of Cnidas, whom he often quotes. A Greek by birth, Ctesias served as a physician in the Persian court for seventeen years and attended Artaxerxes on the battlefield. His history of Persia to 397 BC, written in twenty-three books, survives today only in fragments. Diodorus (or Ctesias) claims that the young Semiramis, nurtured by doves as a baby and brought up by the keeper of the royal herds of cattle, 'far surpassed all the other maidens in beauty' when she came of age to marry. She married an army officer, but unfortunately for him the king, Ninus, accredited by Diodorus as the founder of Nineveh, became infatuated with her and when her husband refused to give her up, threatened his well-being to the extent that he hanged himself. Ninus then married her. Shortly afterwards he died, whereupon Semiramis erected a huge mound over his tomb, then set about founding the city of Babylon known to the classical writers, putting in hand stupendous building projects.

She decided to install a very large obelisk within the city to serve as a focal point and, for this purpose, according to Diodorus, 'quarried out a stone from the mountains of Armenia which was 40m long and 7.5m wide and thick; and this she hauled by means of many multitudes of yokes of mules and oxen to the river and there loaded it on a raft, on which she brought it down the stream to Babylonia'. Such a stone would have weighed well over 5,000 tons, many times bigger than any obelisk raised and transported by the Egyptians, and could not have been transported in the manner described by Diodorus. It served, however, as a spur to Layard, a somewhat eccentric Englishman, who discovered in 1845 the ruins of Nineveh with its bas-reliefs and huge sculptures of human-headed winged bulls and lions, weighing about 10 tons, which he wanted to remove from the site and ship to London. Well versed in the writings of Diodorus and the supposed feats of Semiramis, Layard was not to be deterred by instructions from the Museum of London to leave the statues in place, covered with earth. He moved the statues out of their trenches on greased rollers and lowered them onto robust wooden carts with solid wooden wheels, which were specially constructed for the purpose. He then conveyed them to the Tigris River, where they were loaded onto enormous rafts, each consisting of six hundred inflated sheep and goat skins, and taken down river to Basra and shipped to London.

Herodotus' claims for Semiramis are much more modest than those of Diodorus, referring only to some embankment works to control flooding. He attributes much more major earthworks to a later, entirely legendary, Queen Nitocris, including channel and basin excavations and diversion of the Euphrates to reduce the speed of the current, and thereby creating a devious course to discourage an influx of Medes into Babylon.

There can be little doubt that earthworks – excavations and embankment constructions – were made by the early rulers of Babylon to control flooding of the city and surrounding areas. Unfortunately for the Babylonians, the river, without which the city could not have existed, could also be exploited by their enemies in their assaults on it, the Assyrians in the seventh century BC and the Persians in the sixth century BC taking full advantage of this.

Assyria became an important power in the region around the middle of the fourteenth century BC, although its capital Assur, exploiting its location on the Tigris, had been an important trading post for at least 1,000 years before this, with much of the north–south trade such as copper from Anatolia and tin and textiles from Mesopotamia funnelling through it. Donkey caravans headed eastwards from Assur. Their kings now corresponded on equal terms with the Great Kings of the Hittites and the pharaohs of Egypt, and, while close ties were maintained with the Kassites in Babylon, these sometimes led to military conflicts between the two. Having assumed dominion over all of northern Mesopotamia by 1250 BC, they turned their attentions southwards towards Akkad and Sumer, and in 1250 BC captured Babylon, the king Tukulti-Ninurta recording: 'I captured Babylon's king and trod his proud neck as if it were a footstool . . . thus I became lord of all Sumer and Akkad . . .'. Their occupation of Babylon lasted only eight years, after which they exercised control over southern Mesopotamia only to the extent required to protect their trading and political interests. Over 300 years passed before they resumed their military conquests, annexing south-eastern Anatolia early in the ninth century BC and overrunning Syria to give them direct access to the Mediterranean.

During his reign Assurnasirpal moved the capital of Assyria to the more central location of Kalhu (modern Nimrud). Tablets found there show the Assyrian-controlled territories to have been divided into provincial units, each with a governor responsible to the king and sometimes a member of the king's family. In the seventh century BC the Babylonians, now predominantly Chaldeans originating from tribal settlements along the lower reaches of the Tigris and Euphrates, drove out Sennacherib's appointed King of Babylonia, his own son Ashurnadishum, presumably in the belief that they would be able to withstand any assault the Assyrians could launch against them. But they reckoned without the technological and military genius of this great Assyrian king. He sacked the city in 689 BC, laid waste to it and massacred the people. Directing water from the Euphrates through the city, he left it a wilderness, and as an added humiliation he removed the statue of Marduk to Assyria. With this accomplished, he transferred his own capital from Khorsabad (briefly the capital under Sennacherib's father Sargon II) to Nineveh.

But Babylon was just biding its time. When its retaliation against Assyria came, it was devastating. Forming an alliance with Scythians and Medes, the Babylonians conquered Nineveh in 612 BC and razed it to the ground; unlike Babylon itself, it was never to rise again. With Assyria consigned to oblivion, Babylon entered a new golden age under the Chaldean kings Nabopolassar and his son Nebuchadnezzar, the latter ruling forty-three years from 605 BC. The city now achieved its greatest size and splendour. Although captured by the Persian King Cyrus in 539 BC, it remained a great city for a further half a century until Xerxes, in putting down an internal rebellion in 482 BC, reduced it to a provincial town. Nevertheless, sufficient of the old city remained to impress Herodotus when he visited it in the middle of the fifth century BC.

Cyrus exploited the river in his attack on the city in 539 BC. Well aware that they would eventually come under attack from the powerful Persian king, who was seemingly unstoppable in the expansion of his empire, the Babylonians had stocked up with sufficient provisions to last many years. As expected, Cyrus invested the city, but as the siege dragged on he or his commanders realised that it would require a change in tactics in order to defeat the city. According to Herodotus:

Then somebody suggested or he [Cyrus] himself thought up the following plan: he stationed part of his force at the point where the Euphrates flows into the city and another contingent at the opposite end where it flows out, with orders to both to force an entrance along the riverbed as soon as they saw that the water was shallow enough. Then, taking with him all his non-combatant troops, he withdrew to the spot where Nitocris had excavated the lake, and proceeded to repeat the operation which the queen had previously performed; by means of a cutting he diverted the river into the lake (which was then a marsh) and in this way so greatly reduced the depth of water in the actual bed of the river that it became fordable, and the Persian army, which had been left at Babylon for the purpose, entered the river, now only deep enough to reach about the middle of a man's thigh, and, making their way along it, got into the town.

This stratagem, also described by Xenophon, enabled the troops to enter the city on a night when the citizens were engaged in dancing and revelries associated with religious festivities. Having taken the city in this bloodless way, Cyrus had no reason to sack the city, and life went on very much as before. According to contemporary accounts, admittedly based on Persian sources, the Babylonians welcomed the replacement of the tyrant Nabonidus, son of Nebuchadnezzar, by the Persian king. Cyrus took up residence in the royal palace; but he respected both the religious and political role of the priests and, most importantly, showed proper respect towards the god Marduk. He allowed trade and commerce to go on as before and, wisely, did not impose swingeing taxes on the city, which could have incited rebellion.

Cyrus may well have learnt something about the technicalities of river diversion from Croesus, the Lydian king he had defeated. Before attacking Persia, Croesus had consulted the Oracle at Delphi and was told that if he did so he would destroy a great empire. With this assurance he marched on Persia. In doing so he had to cross the Halys River. According to Herodotus, he traversed an existing bridge, but he also recounts an existing story that, advised by Thales of Miletus, Croesus had the river split into two fordable channels.

In the event, Croesus crossed the river and laid waste to the land, dispossessing innocent Syrians on the other side of their homes and possessions and even their freedom. After a brief battle with the much larger army of Cyrus, he hastily retreated to his capital at Sardis to drum up support from his allies before mounting a further attack on the Persians. Help never came. Cyrus pursued the Lydian army, and after a siege of fourteen days, stormed Sardis and took Croesus prisoner. And so the oracle was fulfilled: Croesus had indeed destroyed a mighty empire, regrettably his own. Cyrus not only spared Croesus, but also befriended him and sought his advice from time to time on political and military matters.

As readily available land for cultivation became scarce in the Greek world, attention turned to the possibility of draining shallow lakes, swamps and marshes to create fertile land. One such area was Lake Copais, a vast reed swamp some 65km north of Athens. The natural outlets, rock fissures and subterranean tunnels often became blocked, particularly by frequent earthquake activity, causing lake levels to rise and flood surrounding fertile land, while the levels of rivers discharging into it also rose and flooded over their banks. Various attempts were made even as early as Helladic times around 1400 BC to overcome this problem, including intercepting incoming water by canal and conducting it to natural outlets.

In 325 BC the Greek engineer Crates made an ambitious attempt to drain the lake, first of all by clearing earlier drainage channels and tunnels, then by driving a tunnel over a mile long. He had the work well in hand when Alexander the Great's military activities in the area brought it to a halt. The practice adopted by Crates comprised driving the tunnel from the bases of vertical shafts, spaced some 60m apart. This gave many faces on which to work and thereby hastened the driving of the tunnel, but was a method that demanded exacting surveying methods. He also adopted a curving alignment, rather than a straight line, following ground that kept the depths of his vertical shafts to a minimum. The work resumed in modern times, with its eventual completion in 1890. The lakebed is now farmland.

An even more famous drainage tunnel of classical times was driven by the Romans under the direction of freedman Narcissus, secretary to Claudius and the most powerful man in Rome, until Claudius, having disposed of his third wife Messalina, married his politically motivated niece Agrippina, a match of which Narcissus unwisely disapproved. She had him jailed and probably killed, having already murdered Claudius with poisoned mushrooms to ensure the succession of Nero, her son by a previous marriage. Five years after his accession in AD 59, Nero murdered Agrippina to rid himself of her domineering influence.

Claudius commissioned Narcissus to oversee the excavation of a drainage tunnel to reclaim 20,000 hectares of land around Lake Fucino in the Apennines, some 80km east of Rome. It took 30,000 men eleven years to drive the 5.5km-long tunnel through limestone and alluvial strata – difficult even with modern techniques – driving forward the excavation, 2.75m wide and nearly 6m high according to Livy, from working faces at the bottoms of forty vertical shafts, up to 120m deep, supplemented by a number of inclined shafts. Considerable stretches of shafts and tunnel in falling ground had to be temporarily supported with timbers and permanently lined with ashlar masonry. Rock was excavated by chiselling or by chilling heated surfaces with water, causing the rock to crack. Excavated rocks and earth were hauled by windlasses to the surface in copper buckets up the vertical shafts.

Claudius ordered a great naval battle on the lake to celebrate the completion of the tunnel, pitting two fleets of triremes against each other, manned by 19,000 expendable convicts. When the first attempt to drain the lake failed because the tunnel inlet was positioned too high in relation to the lake level, the Emperor ordered the mistake to be corrected and staged a second grand opening with further satisfactory bloodshed. He also ordered a banquet for the great and the good to be set up close to the tunnel outlet, part of which, along with some of the participants, was washed away when the outflow proved to be much greater than expected. This was Narcissus' final undoing, as, according to Suetonius, he argued violently with Agrippina over the mishap, she accusing him of profiting unduly from the work and he impugning her personality and lifestyle. Pliny, who must certainly have witnessed the work, described it as beyond the power of words to describe, the operations imaginable only by those who saw them. It remained the longest tunnel in the world until the opening of the Mount Cenis railroad tunnel in the European Alps in 1876.

The Romans made sporadic attempts to drain the Pontine Marshes by canalisation, but with only limited success. They achieved greater success with land reclamation works in the Po Valley, undertaken for the purpose of settling discharged soldiers, and with drainage works in the marshes of Ravenna. Ravenna became increasingly important in Roman times and even more so after the fall of Rome when Honarius made it his capital.

The Romans did not confine their drainage work to Italy, and it is possible that Car Dyke in England, starting just north of Cambridge and terminating near Lincoln, although primarily for transportation, also served to drain parts of the Cambridgeshire and Lincolnshire fens.

Drainage works in the ancient world did not stop at flood control and land reclamation, but also included urban works, some of which showed a high level of technological skill. The remains of an excellent sewerage system can still be seen at the site of Mohenjodaro in Pakistan, one of the twin capital cities of the Harappan civilisation that flourished along the Indus River between 2500 BC and 1500 BC. Excavations have revealed a well-laid-out rectangular grid of unpaved streets, beneath which ran an elaborate system of corbelled drains constructed with kiln-fired bricks. Waste-water matter discharged from private houses and public buildings was conducted through pottery pipes to the under-street drains, which also collected surface water after rains. The drains led the sewage to soak pits situated outside the populated area. Household rubbish was similarly disposed of by being placed into chutes discharging into bins in the street, which the authorities emptied at regular intervals. Most of the houses had their own brick-lined wells to provide water for cooking and washing.

The Minoans, whose civilisation in Crete was contemporary with that along the Indus, also had outstanding hydraulic engineers. Their palace at Knossos had bathrooms, bathtubs and sanitary facilities flushed by a continuous water flow system. The latrines connected to an extensive drainage system through the palace, utilising tapered baked clay or terracotta pipes, which gave a shooting motion that helped to keep the successive lengths clear of sediment.

Many Minoan ideas were passed on to the Greeks, but this did not include sanitation. The streets of most Greek cities were muddy, filthy alleys with no attempt at drainage: refuse and slops were thrown into the streets, a custom that prevailed in European cities until the eighteenth century. The more thoughtful Greek occupant, mindful of the passer-by, would call out *exito!* before throwing out the slops.

Smaller Roman towns and settlements probably had sanitary conditions little if any better than those practised by the Greeks, but in larger towns the Romans installed efficient drainage and sewerage systems, including under-road drains that can be seen in Pompeii and Herculaneum today. The first drainage system in Rome itself started in the Tarquin period, perhaps 500 BC, with the excavation of an open ditch to drain the valleys between the seven hills into the Tiber. The Forum was drained in this way. Successive generations enlarged and improved this drain, and about 300 BC it was covered with a stone barrel vault, sufficiently large to be navigable by small boats. This drain, the Cloaca Maxima, still carries rainwater off the streets into the Tiber, as it has done for more than 2,000 years.

Although the Cloaca Maxima itself has never carried household wastes, the sewerage system created by the Romans for their capital did cope with this problem. A number of large public latrines connected to this system, using water from the public baths and industrial establishments to flush the appliances. Some of these latrines became so elaborate that archaeologists excavating one in the nineteenth century mistook it for a temple. Most of the *insulae*, which were similar to our modern blocks of flats, had their own latrines connecting to the sewer system or to cesspools.

Pit leading to under-road drains in Herculaneum.

Control of water could be exercised in a positive way by harnessing the energy of water flow to provide power. Although there exists no evidence that civilisations pre-dating the classical period in Europe knew about or used waterwheels, it does not necessarily exclude the possibility that they did so. The first reference to a waterwheel appears to be in a poem by Antipater of Thessalonica, in the first century BC, which runs:

> Women who toil at the querns, cease now your grinding;
> Sleep late though the crowing of cocks announces the dawn.
> Your task is now for the nymphs, by command of Demeter,
> And leaping down on the top of the wheel, they turn it,
> Axle and whirling spokes together revolving and causing
> The heavy and hollow Nysrian stones to grind above.
> So shall we taste the joys of the golden age
> And feast on Demeter's gifts without ransom of labour

Strabo, writing in 24 BC, refers to a waterwheel at Cabeira in the Pontus that formed part of the property Methracles lost in 65 BC, when he was overthrown by Pompey.

Surprisingly, the Greeks and Romans used waterwheels only for grinding corn, ignoring other possible applications of the power the wheels provided. It must have occurred to their engineers that the power could be usefully employed for mine drainage, fulling cloth, sawing wood and other purposes, but the availability of other labour may have made such uses unnecessary. It might also have been official policy to keep this abundant labour fully occupied. Idle hands allied to fertile brains might have led to activities not conducive, in the minds of the authorities, to good order in society.

The most primitive type of mill comprised a horizontal wheel with six to eight flaps or scoops fixed around its rim and set in a rapidly running stream, or with a stream of water from a chute directed onto the flaps; in the latter case it became a primitive turbine. The vertical shaft passed up through a hole in the lower millstone and turned the upper millstone, and so did not require any gearing. This type of mill, usually known as a 'Norse Mill' or 'Vertical Water Mill', was common in hilly regions in the Near East. Slow and inefficient, it could grind only small amounts of corn, making it suitable only for single-family or small-community use.

A better arrangement consisted of a waterwheel set vertically on a horizontal axis, driving gearing that not only converted the horizontal rotary motion into rotary motion about a vertical axis, but also allowed the rate of rotation to be greatly increased. Vitruvius gives an excellent description of this type, having already dealt with the raising of water by human treadmill:

Wheels on the principles that have been described above are also constructed in rivers. Around their faces flatboards are fixed, which, on being struck by the current of the river, make the wheel turn as they move, and thus, by raising the water in the boxes and bringing it to the top, they accomplish the necessary work through being turned by the mere impulse of the river, without any treading on the part of the workmen.

Water mills are turned on the same principle. Everything is the same in them, except that a drum with teeth is fixed into one end of the axle. It is set vertically on its edge, and turns in the same plane with the wheel. Next to this larger drum there is a smaller one, also with teeth, but set horizontally, and this is attached (to the millstone). Thus the teeth of the drum, which is fixed to the axle, make the teeth of the horizontal drum move, and cause the mill to turn. A hopper, hanging over this contrivance, supplies the mill with corn, and meal is produced by this same revolution.

The increase in rotation rate through the gearing in Roman mills, commonly about fivefold, provided the fast rotary motion required to grind large quantities of corn. Nevertheless, this type of wheel, the undershot wheel as described by Vitruvius, still had a low efficiency, probably no more than 20 per cent, in addition to which, in deriving its power solely from the velocity of the water striking the underside of the wheel, it suffered from the vagaries of stream flow rate and water level.

Despite the clear description of the water mill given by Vitruvius, they were not in widespread use at the time of his writing in the first century BC and started to find favour only in the fourth century AD, and then apparently only because of an acute shortage of labour. In Rome itself milling activities centred on the Janiculum, using horses and donkeys to drive the mills. In some places, the Roman authorities favoured human-driven

The Roman flour mill at Barbegal, near Arles in France, consisted of sixteen overshot wheels with a total fall of 20m. Driven by water brought to the site by a 9km-long aqueduct, the mill probably produced, on average, about 4.5 tonnes of flour a day.

Sagui's reconstruction of the mill at Barbegal.

mills, exploiting convicts, prisoners of war or simply the unemployed, as a means of absorbing this potentially explosive human energy.

A much more efficient unit, known as the overshot wheel, had scoops or buckets around its circumference driven by a stream of water directed onto the top of the wheel. The energy driving the wheel came from both the velocity of the water and its head or elevation, the weight of water in the descending buckets helping to rotate the wheel. Mills of this type had efficiencies as high as 60–70 per cent and a wheel of typically 2m diameter developed up to 3 horsepower. The drawback with this type of wheel was the need to bring the water in at a high level. This could be accomplished by locating the mill beside a stream tapped at a point some distance upstream and having the water brought to the mill along a chute with a flatter slope and perhaps more direct route than the stream itself. Alternatively, where the topography allowed, the mill could be sited on a hillside and the water brought in by aqueduct to the top of the hill.

Evidence of the use of overshot wheels is fairly sparse, but the remains of such a wheel operating in the Athenian Agora (marketplace, place of assembly) in the fifth century AD have been found. Easily the most impressive find, however, has been the remains of a Roman mill at Barbegal, near Arles in the south of France, built in the fourth century AD. It consisted of sixteen 2.1m-diameter overshot wheels, in pairs, sited on a hillside with a slope of 30° and a total fall of 20m. Water was brought by a 9km-long aqueduct to the site to drive the wheels. Allowing for maintenance and other interruptions, it has been calculated that the mill probably produced, on average, some 4.5 tons of flour per day, sufficient for a population of 12,500, estimated to be the size of the garrison town of Arles at that time. Remains of this mill can still be seen.

Waterwheels may have been in use in China as early as the fifth century BC, and by the third and fourth centuries BC were abundant. Furthermore their use was not restricted to milling corn almost exclusively, as in areas under Roman influence, but extended to a number of operations such as working bellows for iron smelting and working trip hammers for a variety of purposes such as iron forges, hulling rice and crushing minerals. Most common was the horizontal wheel, activated by a tangential jet of water, and thus a direct ancestor of the modern turbine.

The settled, urbanised, way of life became possible in Mesopotamia and Egypt as a result of effective irrigation systems and the cultivation of cereals. The great rivers – the Euphrates, Tigris and Nile – provided the water to irrigate the productive soils of the Mesopotamian plains and the fertile strip fringing the Nile, and wheat and barley, which were already being cultivated in Neolithic times, provided the essential cereals. Conditions that favoured the cultivation of crops also favoured the domestication and breeding of animals, such as goats and sheep, which could provide food and the raw material for clothing and other uses. Cattle also provided these commodities and could additionally be used as work animals. Donkeys and horses were bred to be beasts of burden and to be used for transportation. It was the need to keep a record of these crops and animals that led to the development of writing: owing to the durable nature of the hardened clay tablets, many such records have survived.

Two commonly occurring groups of wheat grew wild, *Einkorn* (with 2 sets of 7 chromosomes) and *Emmer* (with 4 sets of 7 chromosomes), but *Emmer* became the more widely cultivated, evidence of its use dating back to at least 5000 BC. The early harvesting of wild wheat by sickles or reaping knives – sharp flint flakes set in a wood or bone shaft – not only represented an important technological development in itself, but also had an important genetic influence on the cultivated wheat. Most wild wheat when harvested in this way shatters – that is, the seeds fall to the ground. Some mutants, however, did not shatter, and when planting and cultivation of wheat displaced the harvesting of wild wheat, it was these non-shattering seeds that became available for planting. Evidence for this has been found at ancient sites such as Jericho, close to where the wild forms of wheat still grow. Further evolution, combining diverse genetic materials from different species, has enabled wheat to adapt to different habitats. Barley has undergone a similar evolution, and other vegetables have also been domesticated.

A mace head in the Ashmolean Museum in Oxford shows a pre-dynastic Egyptian king, called King Scorpion because he is portrayed with a scorpion hovering in the air in front of him, apparently presiding at a ceremony marking the commissioning of an important irrigation canal. Wearing the tall white crown of Upper Egypt, he is pictured standing on the bank of the canal, hoe in hand and three times larger than the men around him, projecting authority and power. Two of the smaller figures appear to be offering him seeds, while others are working on the canal banks, a nearby reed mat possibly serving to strengthen the banks or even to function as a sluice gate. There was probably a religious meaning to the ceremony, too, reflecting the divinity of the king and the sacredness of the Nile, which provided the life-giving waters.

Unlike the Tigris and Euphrates, the Nile flooded rhythmically as a predictable event, and from August to early October proved most propitious to the growing of crops. This negated any need to build water-storage structures. Irrigation could be effected simply by controlling the floodwaters which spilled over the riverbanks flooding the adjacent river plains. This is known as basin irrigation. The method of control consisted of dividing the land along the river margins into basins by a rectangular grid of earthen bunds parallel, and at right angles, to the river, enclosing areas of 800 to 16,000 hectares. Floodwaters were first led into selected basins to a depth of 1–2m, allowed to soak for a month or so, and the excess water then drained off through sluices into adjacent lower-lying basins. The excess water in the final, and largest, basin was drained back to the river along a canal when

The Egyptian Narmer macehead, now in the Ashmolean Museum in Oxford, shows predynastic 'King Skorpion', hoe in hand, directing operations. Smaller figures include farmers offering seeds and workmen constructing or repairing a canal.

Widely used in ancient times throughout Egypt and the Levant, the shaduf works using the simple principle of a balanced beam with a weight, often just a lump of clay, at one end to counter the weight of the water-filled container. Shadufs are still used today to raise water for irrigation.

The Archimedes screw, in use long before the time of Archimedes, is still used today, as here in Egypt to raise water for irrigation.

the river level had dropped sufficiently. The seeds germinated easily in the mild winter climate, and grain crops were ready to harvest by mid-April.

The Egyptians practised some perennial irrigation, particularly for the watering of smaller vegetable plots. Water was lifted from the Nile and fed into channels supplying the plots, using shadufs or wheels of pots. The former, still used today, consists of a long, balanced wooden beam pivoted on top of an upright timber frame, often consisting simply of two posts and a cross-piece; a vertical rope attached to a water container is fixed to one end of the beam and is counterbalanced at the other end by a large stone or a wodge of dried clay. By hauling on the rope the operator lowers the water container into the river or canal until it is filled, then with minimum effort raises the counterbalanced filled container and empties it into the irrigation channel.

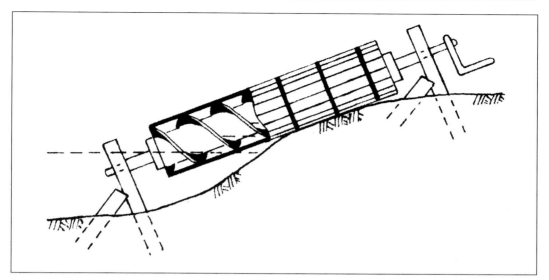

The Archimedes screw.

Hydraulic works for irrigation reached their apogee in Ancient Egypt in the Middle Kingdom beginning about 2040 BC. The greatest pharaoh of the period, Amenemhet III, ruled for nearly fifty years, a period of social solidity and high intellectual attainment. A major canal (now known as the Bahr Yusuf – 'Joseph's arm') was constructed, possibly by re-opening an old branch of the Nile, to Lake Moeris (now Lake Qurun) in the Fayum, a large depression situated to the west of the Nile, which Herodotus thought, incorrectly, to have been dug by man and the excavated soil thrown into the Nile to be carried away.

Lake Qurun today has a surface area of 230 square kilometres, but it is only a small remnant of the ancient Lake Moeris.

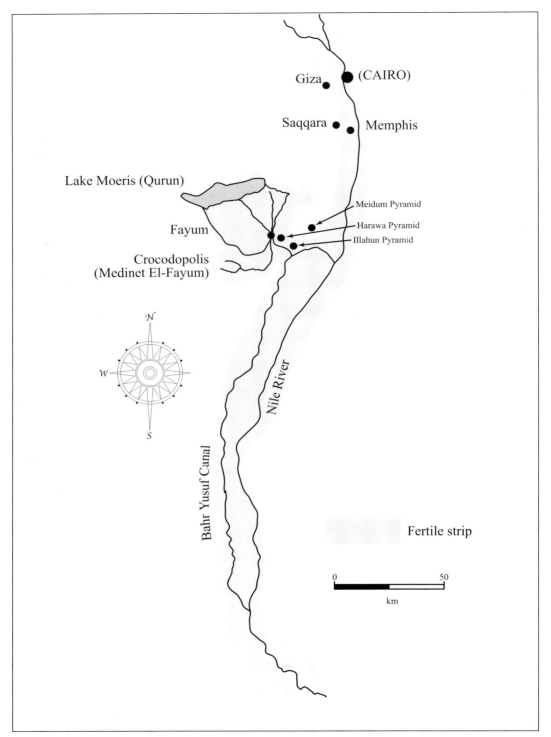

Bahr Yusuf and the Fayum.

The Bahr Yusuf has conveyed water from the Nile to the Fayum for the past 4,000 years.

The Bahr Yusuf Canal leaves the present course of the Nile and runs parallel to it for some 240km before discharging into the Fayum. After passing close to the remains of the mud-brick pyramid of Lahun, it swings to the west to pass close to the remains of the Hawara mud-brick pyramid and splits into a large number of irrigation channels, some reaching Lake Qurun (Birket Qurun). The surface of this lake is 40km below sea level and, although much smaller than 4,000 years ago, it still has a surface area of about 230km². The work carried out by the Middle Kingdom pharaohs, Sesostris II (1897–1878 BC, also known as Senwosret II) and Amenemhet III (1844–1797 BC), consisted of partially draining the lake to increase the area of fertile land, canalise the Bahr Yusuf channel and dig linking channels to enable widespread irrigation throughout the Fayum, and to build structures in the canal and irrigation channels to control and possibly measure the flow of water to the fields. The exact nature of the structures is a matter of some dispute: Strabo refers to them as *artificial barriers* and Diodorus describes them as both skilful and expensive. This may suggest some sort of weir structure, which could be raised or lowered to allow and control the flow of water and, by measurement of the depth of water over the weir, give a measure of volumetric rate of flow.

The dedication of the town of Shebet to the crocodile god, and the transfer by Sesostris II of his capital from Memphis to Illahun on the fringe of the Fayum, attests to the importance of the region in Middle Kingdom Egypt. Its importance derived from its

There is a marked contrast between the desert and the irrigated and cultivated area of the Fayum.

extreme fertility, made possible by the enlightened work of the Middle Kingdom pharaohs and their engineers in draining and irrigating the land. Further extensive operations in the Fayum were carried out in Ptolemaic times, which, in addition to drainage and irrigation work, included setting up the ancient town of Philadelphia some 40km north-east of Shebet. It is well known to papyrologists for the large number of mummy portraits (painted on wooden panels or shrouds covering the bodies) and other documents found in its necropolis. The Fayum remains today the garden of Egypt, with its abundant harvests of vegetables and sugar cane, and its groves of citrus fruits, nuts and olives.

In order to discharge their responsibilities, the provincial governors in pharaonic Egypt employed inspectors to oversee works in progress and to exercise control over the usage of the irrigation waters. Inspectors were also sent out by the central authorities to assess crop yields for taxation purposes. This led to the establishment of a civil service structure at local provincial level and at central government level.

Although the priests used astronomical observations to maintain a mystical hold on the people, these observations also served the practical purpose of predicting rhythmical events such as the inundation of the Nile. These predictions were backed by the installation of vertical scales or *nilometers* marked in cubits and fractions of cubits, installed at various locations on walls or quays flanking the Nile. Calculations had to be made for a wide range of purposes: land areas; volumes of water, earthworks and crops; rates and quantities of water flow. This needed some understanding of mathematics, geometry and trigonometry.

Herodotus not only understood well the importance of the great rivers of Egypt and Mesopotamia in providing the water to irrigate these parched lands, but also the different methods employed arising from the differences in the two river systems. He observed, too, that the (cultivatable) black and friable soil of Egypt, formed of the silt brought down the Nile from Ethiopia, differed from the stony and clayey soil of neighbouring Arabia and Syria. He also comments on the crops grown:

> As a grain growing country Assyria is the richest in the world. No attempt is made there to grow figs, grapes or olives or any other fruit trees, but so great is the fertility of the grain fields that they normally produce crops of two-hundredfold, and in an exceptional year as much as three-hundredfold. The blades of wheat and barley are at least three inches [76mm] wide. As for millet and sesame, I will not say to what an astonishing size they grow; though I know well enough; but I also know that people who have not been to Babylonia have refused to believe even what I have said already about its fertility. The only oil these people use is made from sesame; date palms grow everywhere, mostly of the fruit-bearing kind, and the fruit provides them with food, wine, and honey.

The Mesopotamian rivers experience more abrupt rises and falls than the Nile, which makes them much less predictable. Furthermore the two Mesopotamian rivers themselves differ in their relative behaviour. Although carrying only 40 per cent of the flow of the Tigris, the Euphrates provided most of the water used for irrigation, because its higher level above the surrounding plains made it much easier to feed water into the canals accessing the irrigation system. The maximum flow and flooding of the two rivers occurs in a period from April to early June, four months earlier than the Nile, and at a time unsuited to the planting of crops. In consequence, the Mesopotamians had to practise perennial irrigation, requiring both river control and water storage.

Implementation of these schemes again needed skills in mathematics, geometry and trigonometry and, above all, in surveying techniques required to set out the courses of water channels, often many kilometres in length, to ensure adequate water flow as well as minimum weed growth and silt deposition, but not so steep as to cause scour of the channel bed or side slopes. This was a challenging problem in the marly soils of the plains, which eroded easily and charged the canal waters with large amounts of sediment, often leading to silting up of canals. Considerable skill and experience would have been needed to get the canal slopes and geometry right using simple water levels.

The great kings of the time made sure that they associated themselves with the irrigation projects vital to the lives of their subjects. A limestone stele exists recording the many great achievements of Ur-Nammu, the great king of Third Dynasty Ur, not least a list of irrigation canals dug at his instigation. A pictorial representation of the king shows him standing in an attitude of prayer while an angel is depicted pouring water on the ground, demonstrating the close relationship between irrigation/agriculture and religion. Tablets found at the site of Nippur have revealed much about Sumerian religion, as well as a farmers' almanac providing advice on how to farm successfully; it recommends, for instance, a prayer to Ninkilim, the goddess for field mice and suchlike creatures, to protect young seedlings from their depredations.

Baked clay tablet found at Nippur in Mesopotamia, dating from about 1300 BC, showing a loop of the main canal and connecting channels, mostly supplying water to irrigate the adjacent fields. The map shows areas for crops and fields for grazing, perhaps on a rotational basis, as well as marshlands providing reeds for building and a number of townships (circled areas).

Although the kings of Third Dynasty Ur established a governorship in Babylon, showing that they considered it to be a place of some significance, it did not become an important player in the region until the Amorites took it over shortly after 1900 BC and established a dynasty lasting some 300 years. In 1792 BC, about 100 years after the establishment of the Babylonian Amorite Dynasty, Hammurabi became king and was destined to become one of the outstanding names in history, although in the first thirty years of his reign there was little to indicate that he would earn this accolade. He used this period to hone his administrative skills and to turn Babylon from a relatively unimportant town into a city to be reckoned with. In these early years of his reign Babylon was surrounded by powerful potential foes. A powerful Larsa king, Rim-Sin ruled southern Mesopotamia, the Akkadians with their capital at Eshnunna (Tel Asmar in the Diyala Valley, 35km north-east of Baghdad) dominated the area to the north of Babylon, while even further north, in Assur on the Tigris, the Assyrians were beginning to flex their muscles under Shamshi-Adad (1813–1781 BC), a ruler of considerable military and administrative ability. His place in history is ensured by the finding of a large number of tablets alluding to his reign at Mari on the Euphrates 400km north-west of Babylon, an important town and staging post for caravan and river traffic, administered by an Assyrian viceroy. These palace archives from Mari have also revealed much that is known about Hammurabi. After the death of Shamshi-Adad Assyrian power declined and from 1762 BC onward Hammurabi won great victories over Assur, Eshnunna and Mari, having already defeated the Elamite king of Larsa, giving him total control over all of Mesopotamia and Assyria. He styled himself, with some justification, 'King of the Four Quarters of the World'.

By 1760 BC Hammurabi had been king of Babylon for about thirty years, and had brought large areas of Mesopotamia under the hegemony of the city, resulting in the strong centralised control of irrigation. Under Hammurabi's direction the various laws of Sumeria and Babylonia, many of them to do with irrigation, became synthesised into his famous Code of Laws. Punishments for wrongdoers could be severe as, for example, in Section 53: 'If anyone is too lazy to keep his dam in proper condition and does not keep it so; if then the dam breaks and all the fields are flooded then shall he in whose dam the break occurred be sold for money and the money shall replace the corn which he has caused to be ruined.' Clay tablets preserve copies of some of the writings of Hammurabi to his provincial governors, each of whom took responsibility for the upkeep of irrigation works in his own district. In a letter to Sid-Idinnam he complains that a canal to bring water into the city of Erech had not been properly completed and had partly collapsed, and he ordered that the canal be completed and cleared within three days. Other clay tablets from the time of Hammurabi and his successors preserve legal documents concerning litigation about water rights, opening and closing of canals, management of sluices and even about poaching in canal waters.

Considering the Asssyrians' deserved reputation as being among the most ferocious and successful warriors of all time, it is difficult to envisage their capitals as cities of peace and tranquillity, generously provided with supply canals and water channels feeding gardens and orchards. Assurnasirpal (883–859 BC) had a canal excavated from the Upper Zab to his capital al Kalhu (Nimrud) for just this purpose. In 691 BC King Sennacherib commissioned the construction of the 80km-long Jerwan aqueduct to bring water from a tributary of the Greater Zab River and to augment the water supply for his capital at Nineveh. It comprised a prodigious feat of hydraulic engineering and masonry construction, requiring the quarrying, transportation over 16km or more and placing of some two million heavy limestone blocks. On the wall of the gorge, above the river barrage feeding the aqueduct, an inscription states that the work was finished in a year and three months. As the work neared completion Sennacherib sent two priests to the upper end of the canal to perform the proper religious rites at the opening ceremony. It must have tested their faith, not to say their nerve, when a sluice gate failed just before the opening, allowing water from the Atrush River to gush down the channel in advance of the king's command. With political adroitness they quickly turned it to their advantage, interpreting it, for the benefit of the king, as a good sign from the gods, who were anxious to see the canal in operation. On arrival at the site the king ordered its repair, sacrificed some oxen and sheep on the advice of the priests, and much to their relief and surprise rewarded the engineers with rich clothing, jewellery and gold daggers. Hydraulic engineers must surely have enjoyed a high status in this society. Other works commissioned by Sennacherib included dams, the canalisation of streams in the hills above Nineveh, at least one tunnel and a system of canals to transfer water from a mountain stream into the city of Arbil. Sennacherib's son and successor, Esarhaddon, added to these hydraulic works.

In constructing such massive schemes these ancient civil engineers must have become well versed in the skills of masonry and earthworks, stimulating the progressive improvement in techniques and tools. A 16km-long channel commissioned by Sennacherib in 703 BC, to supply water from the River Khosr to augment the flow of the Tigris at Nineveh, was dug with iron pickaxes. Earthen banks, bunds and dams had to be constructed in such a way that they remained stable, and early geotechnical engineers responsible for these structures often devised ingenious techniques for ensuring stability. Reed mats were inserted into earth banks to strengthen them, and placed on dam slopes to protect them from scour. Facings of burnt brick and readily available bitumen were commonly used to protect and stabilise earth structures. Various techniques for water-raising evolved, leading eventually to simple wooden cog-wheels or gear-wheels to give mechanical advantage, or to convert horizontal circular motion imparted by an animal to a useful vertical motion.

As civilisations developed, established themselves, and then disappeared, the one factor that remained constant was the need to maintain and extend the already long-established irrigation systems. As the land in some areas salted up new areas had to be opened up to agriculture to provide the foodstuffs required by the great cities. Ascendant civilisations profited from the knowledge and technology passed down from in, or in some cases wrested from, older civilisations.

The establishment of the Sassanid Dynasty in AD 224 marked a resurgence in Persian dominance in lands from the Euphrates to the Indus. In AD 260 at Edessa, Rome suffered its most disastrous defeat up to that time at the hands of the Sassanian King Shapur I, who captured the Emperor Valerian and took 70,000 prisoners, containing within their numbers many of the outstanding civil engineering skills of the Romans. Shapur put the prisoners to work building a masonry dam over 500m long across the Karun River near Shushtar, having first diverted the river through an artificial cut. Only 2 or 3m high, the dam was surmounted by a masonry bridge.

After the success of this scheme, other similar projects followed, making full use of available Roman expertise and labour. In time, the Sassanians developed their own expertise and became outstanding civil engineers. Irrigation in Sassanian times reached its peak at the time of Chosroes I (AD 531–79), as evidenced by the Nahrwan canal – 450km long – taking off from the left-hand bank of the Tigris some 200km north of Baghdad and rejoining the river a similar distance to the south of that city. Part of a complex irrigation scheme, it captured left hand tributaries flowing into the Tigris, the complete system requiring various dams and control works. The main canal was a huge exercise in excavation, with a width of up to 100m or more and a depth of 8m. It is likely that parts of the Nahrwan scheme were constructed or reconstructed during the period of the Abbasid Caliphs after their defeat of the Sassanians in AD 636 and again in AD 641.

Remains of ancient masonry or masonry/earth barrages can still be seen in a number of locations in the Near East. Some have silted up; others, with inadequate spillway provision, long since breached by overtopping. No doubt military activity accounted for some and even earthquakes may have taken their toll, as they certainly did with many ancient buildings. One of the most impressive barrages, 600m long, closed off a large wadi near Marib in the Yemen. The first structure at the site, of which no trace remains, may have been an earth dam built around 750 BC. This was replaced by an earth structure faced with stone slabs set in lime mortar to render the structure watertight and to resist wave action. It had a symmetrical triangular cross-section with 45° slopes. Originally 7m high and built around 500 BC, this structure underwent additional heightening to 14m, around 100 BC, to cope with the reservoir silting up and the increased demand for irrigation water, as land areas under cultivation increased.

Water was drawn off for irrigation through sluice gates at both the northern and southern ends of the dam. A spillway at the northern end discharged overflow water into a series of channels that took the water away from the dam. Irrigation water drawn off through the double sluice gates at the northern end passed first into a settling tank and then along a stone channel 1km long to a distribution tank, where four sluices on each of three sides led to water distribution channels. The southern sluice gate, situated in a cleft of rock, led to two hewn channels. Masonry walls up to 8m high formed part of the

An ancient barrage 600m long, built around 500 BC, closed off a large wadi near Marib in Yemen. Its primary purpose may have been to raise water to supply irrigation channels. Shown here are the remains of sluice gate and channel structures.

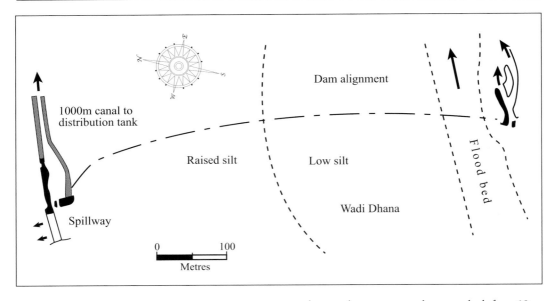

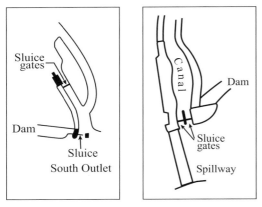

southern abutment and extended for 60m along the channel margins. This masonry work was of the highest quality, with the carefully cut and fitted blocks held together by lead dowels without mortar. The reservoir continued in use, to a greater or lesser extent, for over a thousand years with numerous repairs to the dam. It finally failed for the last time in AD 575.

The vast area of the Indus basin lying mostly in India and Pakistan, extending to more than 1 million km² and with a major river system having an aggregate length of over 5,000km providing abundant water, presented a natural setting for the development of early civilisations corresponding to those in Egypt and Mesopotamia. That such civilisations existed is attested to by the remains of two major twin cities – Harappa and Mohenjodaro – which reached their peak around 2000 BC.

Irrigation in the Indus basin dates back some 4,500 years, the earliest settlers no doubt practising a simple form of inundation by breaching the riverbanks or natural river levees during periods of high flow and flooding narrow strips of land close to the rivers. These methods could not have supported the populations of cities as large as Harappa and Mohenjodaro, the needs for which could only have been satisfied by irrigation systems that involved the raising, storage and canalisation of water. Evidence of such systems, although probably of a later date, survives in the form of the remains of dams constructed of hewn masonry, found in parts of the Sind, Baluchistan and the sub-mountainous regions in the north. These were constructed across watercourses and small streams to store water during the rainy seasons, many of them eventually silting up leaving flat cultivatable fields.

The arrival of Aryans into the Indus Valley around 1500 BC gave a new impetus to the region and saw the expansion of irrigation systems to supply settlements established well away from the rivers. This required long canals of large capacity and the construction of reservoirs or 'tanks' for the storage of irrigation water. It became common to grow more than one crop a year and irrigation was extended to fruit growing. A number of references to irrigation and associated operations, such as the digging of wells, construction of dams and canal excavations, are made in the Vedas, the earliest sacred books of the Aryans, written around 1000 BC. Around 300 BC the Greek ambassador to the court of Emperor Chandragupta Maurya, one Megasthenes, described the duties of the District Officer:

> He measures the land and inspects the sluices by which the water is distributed into the branch canals so that everyone may enjoy his fair share of benefit.

Surviving rock inscriptions attest to the use of masonry dams to store water and attribute such a structure to Emperor Maurya. It was later strengthened in the time of Emperor Ashoka, and lasted for 400 years.

When Sinhalese King Vijaya invaded Sri Lanka in 504 BC, he and his successors proceeded to build one of the most elaborate irrigation systems in the ancient world, which, with additions and extensions, continued for a period of over a thousand years. Two different systems evolved for conserving the seasonal rains of the two monsoons. The first of these consisted of embanking the upper reaches of a valley to create a reservoir fed entirely by water from its own watershed. This limited the amount of water available for storage and led to the adoption of more ambitious schemes of supplying water to reservoirs from outside catchments. Massive masonry barrages known as *anicuts*, thrown across the larger perennial rivers, raised the level of the water, allowing it to be diverted into excavated channels that carried it, in some cases over many kilometres, to be impounded in large tanks. Embanking natural depressions with earthen bunds around the perimeters, in stages, to form a series of such reservoirs along their valleys formed these tanks, often very large in size.

While the tanks served mostly to store irrigation water for the rice fields, some also provided water for communal usages. A number of tanks sited close to, or within, Anuradhapura, the ancient capital of the island, stored water for use in pleasure gardens and for bathing ponds. These tanks, the earliest of which may have dated back to the fifth century BC, received their water from channels and aqueducts fed by rivers to the north of the city. The genius of Sri Lanka's hydraulic engineers did not stop with the raising of river levels by the construction of masonry *anicuts*, the excavation of channels and the building of earthen bunds. They also invented the *bisokotuwa*, or water access tower, constructed within the reservoir, and the masonry sluice, or *sorowwa*, to carry water from the reservoir under the bunds to the feeder channels.

Reputedly the most beautiful lake in Sri Lanka, Kantalai tank, dating from the seventh century, occupies a 1,500-hectare site near the north-eastern port of Trincomalee. A nineteenth-century traveller described his experiences in coming across this expanse of water:

After several hours of travelling through the dense forest, it is with a shock of delight that the monotony is broken by the sudden appearance of a beautiful lake stretching away for miles to dreamy ranges of distant hills, whose beauties are reflected in its calm waters. Light and life combine to treat us as we emerge from the dense jungle. Flashes of every tint appear as the gay birds are startled by our approach. We stand enchanted by the scheme. All is still save the voices of the creatures that dwell on these beautiful inland shores.

The creatures he saw included 'spotted deer, peacocks airing their gaudy plumage, herds of buffaloes, grim crocodiles, troops of chattering monkeys, stately cranes and pink flamingos'.

Although formed largely in a natural depression, the bunding required to close the lake included a major earthen structure 2.5km long, 15m high and 60m wide at the base. Various restorations and other works since the middle of the nineteenth century have increased the capacity of the tank, enabling it to provide sufficient water to irrigate 7,000 hectares of prime agricultural land. A breach occurred in the bund in 1986, at the location of a sluice added in the nineteenth century, releasing water from the reservoir and claiming the lives of at least one hundred people as well as leaving thousands homeless.

Despite huge extensions in the twelfth century, at the instigation of King Parakramabahu, the entire irrigation system had become neglected and derelict. By the time the Portuguese had established themselves in Sri Lanka in the sixteenth century the jungle had claimed both Anuradhapura and the later capital of Polonnaruwa. The people had deserted the plains for the mountains, possibly as a result of Tamil invasions from southern India, but, perhaps more likely, to escape the widespread ravages of malaria. The abundance of stagnant water, resulting from its copious and no doubt sometimes wasteful use in irrigation, and the numerous village tanks, would have presented ideal breeding grounds for the anopheles mosquito. It is unlikely that the role of the mosquito in spreading this disease was understood, but the people found they could escape the disease by retreating to higher ground.

The Chinese introduced perennial systems of irrigation at least 5,000 years ago, using earth-bunded surface tanks and reservoirs to store water. Wet rice cultivation needed not only large quantities of water, but also a highly developed regulatory system to flood the paddies to just the right depth, and also to drain them and re-flood them during the growing period. By the time of the Han Dynasty (206 BC–AD 220) intensive paddy rice cultivation had become widespread in southern China, and the farmers showed considerable enterprise in selecting fine ears for breeding better strains of rice. The technique of multiple rice/wheat cropping appears to have been introduced during the time of Empress Wu (AD 650–704) of the Tang Dynasty. Rice nevertheless retained a dominant position amongst cereals, with vast quantities of it being shipped from the south to the north of the country along the Grand Canal. The demands of the north for more rice led to the practice of growing two or three crops annually. Ironically, during the long periods of feudal rule in China, only the rich and middle class could afford to eat rice daily, and those who grew the crops lived mostly on wheat, barley, potatoes and taros.

Two of the most outstanding and enduring irrigation projects in Chinese history date from the Ch'in period. Both are still in use in modified and extended form. One of these, the Chengkuo canal completed in 246 BC, marked the start of a long series of major projects for the irrigation of the Wei Valley, which had the purpose of providing not only water, but silt as well to replenish the soil annually. An early account of this project describes how the Han people tricked the State of Ch'in into undertaking this major work in the belief that it would exhaust the Ch'in and make them vulnerable to military attack. To this end they even sent Cheng Kuo, a hydraulic engineer, to the Ch'in to direct the work. Although the Ch'in became aware of the trick, Cheng Kuo managed to persuade them that the project would be greatly to their benefit, which indeed proved to be the case as the harvests from their fertile lands made them sufficiently rich and powerful to defeat all other feudal states militarily. The scheme consisted, essentially, of excavating a canal 150km long from the Ching River to irrigate an area of 250,000 hectares. Many modifications, additions and remedial measures have been made to this project up to the present day; in particular it has been necessary several times to re-cut the main canal as it has silted up, each new excavation requiring the construction of a new intake higher up the Ching River. The intake is now high up in the rocky defiles of the river and the works include a 400m-long tunnel under a large dam. The canal also traverses a number of streams by means of eleven bridges. In AD 995, 176 sluice gates were introduced into the system.

Needham draws a comparison in scale between the second major Ch'in project, the Du Jiang Yan (Kuanhsien) irrigation system, and the ancient works of the Nile. As with the Cheng Kuo canal, the economic benefits of this scheme added strength to the Ch'in and helped cement their hold on the country. The basis of the scheme comprised a stone embankment in the form of a snout pointing upstream which Li Ping, the Governor of Szechuan, directed his engineers to construct in the water course of the Min River. This divided the flow into two channels, the outer one following the old course of the river and the inner one taking a new course skirting the city of Kuanhsien, which required an excavation known as the Cornucopia Cut, 27.5m wide and up to 40m deep through hard, conglomerate rock. As well as supplying irrigation water, the outer channel carried some boat traffic and also, during flood times, took excess water from the inner channel, which flowed over a spillway set in the wall of the inner channel, to maintain its flow at optimum level for its sole purpose of providing water for irrigation. Water from this inner channel debouched into laterals and sub-laterals after passing the city of Kuanhsien. Today it totals over 1,200km in length and is capable of delivering water to over a million hectares of land.

Maintenance of the system is carried out annually and is based on the recommendations set down by Li Ping. To do this it is necessary to close off the outer channel with a temporary coffer-dam in mid-October, when the river flow is declining, allowing it to be dredged and repairs made if necessary and then, at low water in mid-February, switching the coffer-dam to the inner channel to give it the same attention. Removal of the coffer-dam in early April marks the start of the irrigation season and is an occasion for celebration. A delightful inscription on a stone in the nearby temple of Erh-Lang, the son of Li Ping, who completed the work after the death of his father, describes the operations for good maintenance of the system:

Water control for the Du Jiang Yan (Kuanhsien) irrigation system included division of the Min River into two channels, the inner on the right as seen here and the outer on the left. Bridges of traditional suspension form connect the island dividing the river to the outer banks of the channels.

Dig the channel deep,
And keep the spillways low;
This Six-Character Teaching
Holds good for a thousand autumns.
Dredge out the river's stones
And pile them on the embankments,
Cut masonry to form 'fish-snouts',
Place in position the 'sheep-folds',
Arrange rightly the spillways,
Maintain the overflow pipes in the small dams.
Let the *(bamboo)* baskets be tightly woven,
Let the stones be packed firmly within them.
Divide *(the waters)* in the four-to-six proportion,
Standardise the levels of high and low water
By the marks made on the measuring scales;
And to obviate floods and all disasters
Year by year dredge out the bottom
Till the iron bars clearly appear.
Respect the ancient system
And do not lightly modify it.

This description probably dates from about the thirteenth century, but is clearly describing long-established practice. 'Sheep-folds' were cylindrical gabions of wooden slats containing stones. The four-to-six proportion describes the controlled division of flow between the inner and outer channels, the inner carrying 60 per cent of the flow at low water, and the outer carrying 60 per cent at high water. The iron bars referred to were a set of three, some 3–4m long and weighing two-thirds of a ton each, laid in the base of the channel to indicate the correct depth for dredging.

The narrow strip of land separating the Peruvian Andes from the Pacific coast is one of the most arid regions on earth, but is traversed by a number of short rivers carrying the run-off from the western slopes of the Andes to the sea. In their upper regions these rivers are steep and are contained within narrow, precipitously sided valleys cut down by their fast flowing waters. On meeting the flood plain the river regime changes to assume a more sedate passage through flat lands with alluvial deposits laid down in the past by the rivers themselves. Each alluvial delta is separated from the next one along the coast by desert sands, which form the sides of the flat coastal valleys.

Small villages grew at the mouths of these rivers to become recognisable settlements by 2500 BC, evolving, with time, into microcosmic civilisations generally isolated and mistrustful of each other. The earliest people lived mainly by fishing and raising a few crops such as beans and potatoes, which grew wild in the Andes. Gradually crops such as manioc and pepinos, as well as potatoes and beans, replaced fish as the staple diet and simple irrigation systems were employed to nourish these, consisting basically of flooding fields close to the rivers by breaching banks or levees at times of high water. Around 1350 BC the introduction of corn, by people emigrating from central America, spurred the improvement of irrigation techniques, as the corn had to be grown on the alluvial soils away from the rivers rather than in the waterlogged soils immediately adjacent to them. But the corn needed adequate water at suitable times, which could not be provided by simple flood farming. Showing commendable resourcefulness, the villagers learnt to tap water from the rivers upstream, at such a level that the water could be fed into small canals, to be conducted to the coastal valley sides away from the rivers.

Irrigation based on the River Moche is typical. From its source in the Cordillera Blanca, the Moche plunges down a steep mountainous slope for the first 85km of its course to meet the coastal flood plain 25km from the sea. It is a perennial river with high flows between January and March and low flows between July and October. Canal irrigation based on this river commenced in the Chavin period (1000–200 BC) and was gradually extended over the next 1,000 years on the north and south sides of the river, but without any apparent central authority until the late Mochica period (c. AD 1000), which saw the construction of major aqueducts and accelerated extension of irrigated field systems. This momentum continued through the Imperial Chimu (AD 1250–1462) and Inca (AD 1462–1532) periods. During the Chimu period the settlements along the coast formed a loose federation of city states with their capital at Chan Chan, a magnificent city in the Moche Valley, with a population of over 100,000 housed in adobe brick buildings, served by streets, reservoirs and supply channels, temples and storehouses. On assimilating the city into their empire in 1462, the Inca not only preserved the irrigation system, but also continued to extend it.

The main offtake canal from the River Moche, the Vichansao, sourced its water from the river at a level 175m above sea level and, after a short steep initial section, it followed the desert margins north of the river at an average gradient of 1 in 2,400, requiring extensive embanking and cutting. Probably started around AD 250, it had grown to a length of 28km by the beginning of the Imperial Chimu period, supplying water to over 2,000 hectares of fields; extensions in the Chimu period increased its length by 7km and added another 1,000 hectares to the irrigated area. For most of its length the channel had a width of about 2m and a depth of 1–1.5m, with dry cobble banks to protect it from erosion. Despite work by engineers in the late Chimu–early Inca periods to straighten various meanders in the canal and improve its flow, it could not cope with the increasing demands placed upon it, precipitating the bold decision to bring water from the Chicama River to the north, a larger stream with a more regular flow than the Moche. This was accomplished by extending an existing internal Chicama Valley canal across the watershed separating it from the Moche Valley to link up with the Vichansao, giving the Inter-Valley canal a final total length of 79km.

Away from the coastal strip, terrace irrigation was practised, although on a scale much smaller than the coastal schemes. This system, parts of which are still in use today, probably extends back as far as the earliest periods of coastal irrigation. Terrace irrigation demanded much less water than the coastal schemes, partly because of the shortage of suitable, gently sloping land for this purpose in the generally rugged terrain of the Andes, and partly because the highlands experienced a wet season. Irrigation played a supplementary role to the rainfall in ensuring adequate water throughout the growing season, and perhaps in extending this season.

The terraces consisted of narrow benches faced with drystone walls commonly about 2m high, but in some cases as high as 4m. Water had to be brought into the system at the level of the top terrace, which meant relying on springs or a high-level water source, often some distance, perhaps several kilometres, away, requiring it to be brought to the site by means of an open channel along a falling contour. Farrington describes such a channel, the Quishvarpata in Cuzco Province, as having a rectangular section up to 800mm wide and 300mm deep, with granite blocks lining the floor and sides to protect it from erosion. To cope with the total drop of 800m in the 6km length of the channel, three steep chutes, each some 50m high, had to be constructed near the entrance to the field system to effect a rapid drop in elevation of the water and, at the same time, dissipate its energy.

Water reaching the front of the top terrace spilt over the drystone wall onto the terrace below, probably into a lined channel at the base of the wall to avoid erosion. This, in turn, distributed water onto the second terrace from which excess water spilt onto the third terrace, a process that continued through all terraces. Excess water from the lowest terrace probably found its way into a local stream.

When the Spanish invaded Peru in the sixteenth century they found, according to Spanish writer Cieza de Leon, 'a veritable garden'. He went on to record: 'Only land that could be irrigated was under cultivation, but this ground was watered and manured like a garden. Corn was the principal crop, reaped twice a year and yielding abundantly. Also grown were yuccas, many kinds of potato, beans, pepinos, guavas, avocados, star-apple and cotton.' Although the Spanish Council of the Indies in far-away Madrid ordered that

Inca irrigation terraces still in use at Pisac, near Cuzco in Peru.

roadworks and irrigation were to be left undamaged, the Spanish invaders took little notice of this in pursuing dominance and wealth and ignored, too, the Council's legislation banning the slavery of the Indians and their use as load carriers and unpaid labour. Huge deposits of silver found in Bolivia in 1545 spelt the end of the unique Indian civilisation and most of the irrigation schemes fell into disuse as indigenous peoples were brought from all over the Andes to work the mines, where huge numbers of them died.

In the early developments of various civilisations, irrigation for crop production transcended all other uses for water. In any case, establishing settlements for the most part close to major rivers and streams ensured a ready supply of water for drinking and for personal usage such as bathing and the washing of clothes and other articles. If the river water was not suitable for drinking, catching rainwater and directing it into suitable receptacles could provide an alternative source of clean water. Otherwise, it could be obtained from sources such as springs or by putting down wells to tap the water table. Jericho, which may have been first settled as early as 8000 BC, owed its existence to a permanent spring that ensured a reliable source of water. It still provides water for Jericho today at 1,000 gallons per minute.

Although early dam construction was almost invariably for purposes of irrigation, either to raise water levels or to raise and store water, the earliest known dam appears to have been constructed to provide water for industrial use. The Sadd-el-Kafara dam built near Helwan, about 30km south of Cairo, sometime around 2850 BC, provided water for a community engaged in mining alabaster. Most of the water would have been used in the mining operations, but some may have been used to satisfy the community's personal needs and perhaps even to water vegetable plots. A remarkably large dam, it had a crest length exceeding 100m and a height of 11m. The cross-section consisted of a core of loosely placed gravel and stones supported by rubble-masonry outer zones, with limestone-block facings for protection against erosion and wave action. It clearly had poor water-retaining abilities and may have failed as a result of water passing through it rather than by overtopping.

Ancient hilltop cities and fortresses invariably faced a problem of water supply within the walls, particularly during times of siege. One solution consisted of sinking shafts or cisterns down through the rock to the water table, but where this was impractical other sources had to be exploited. Sometime before 1000 BC, the Tebusites constructed a tunnel from the spring of Gibon, outside the walls of Jerusalem, to a cistern under the city, access to which was by a 13m shaft to the surface. One of David's warriors, Joab, led a party of Israelites through the tunnel and captured the city. David, King of the Jews, made Jerusalem his capital and Joab Commander in Chief of the Army. A similar system to that in operation in Jerusalem supplied water to the hilltop citadel of Mycenae in the fifteenth century BC. Underground spring waters were collected in a cistern accessed by a stepped passage, which descended through the rock undergoing several changes of direction.

Where water could not be obtained from surface sources or by sinking vertical wells, other sources of supply had to be sought. One solution, dating perhaps from 1000 BC or even earlier, was the *qanat*, supplying water for both irrigation and domestic use. When the Assyrian King Sargon II invaded Armenia in 714 BC, he saw this system in operation and his son, Sennacherib, made good use of it to augment the already copious water supplies of Nineveh and other Assyrian cities. *Qanats* are still constructed today.

A *qanat* is a gradually sloping tunnel tapping an underground water source in the hills and conveying the water under ground to the plains at the foot of the hills, where it discharges into water retention basins or into a system of irrigation channels. The tunnel slope is usually between 1 in 500 and 1 in 1,500. It is constructed by sinking a line of vertical shafts, some 50m apart, along its projected line, then connecting the bottoms of the shafts by a continuous tunnel, a not inconsiderable exercise in surveying. The vertical shafts serve as ventilation ducts and also for the removal of spoil from the tunnel excavation. The technique became common throughout areas that came under ancient Persian influence, from Pakistan and Turkestan to the Yemen, and under later Roman and Islamic influence spread to North Africa, Spain and Sicily. Many of these systems are still in use today, most notably in Iran, where 50,000km of these underground conduits provide up to three-quarters of all water used in that country, including irrigation water.

Although the Romans are justifiably renowned for their water supply systems in Rome and the provinces, they may well have learnt much from Greek and Etruscan engineers, who also displayed considerable talents in this field. An outstanding Greek example was

Stairway to the underground water supply cistern at Mycenae.

the water supply for Pergamon, built during the time of Eumenes II around 200 BC. The water flowed by aqueduct from a spring high up in the mountains to settling tanks at a level of 360m above sea level, whence it flowed by pressure line into the city at a level of 332m. The pressure line traversed two valleys separated by a ridge, acting, in effect, as an inverted siphon. Maximum pressure in the lead pipeline reached 18–20 atmospheres, but, not surprisingly as lead is a valuable metal with many uses, none survives to the present day. Chemical analysis of the soil supporting the pipeline has confirmed the former presence of lead. Evidence for the pipeline survives in the form of numerous stone anchor blocks with holes to receive the pipes.

An even more remarkable Greek achievement was the tunnel commissioned by Policrates and constructed by the engineer Eupalinus around 525 BC, to supply water for Samos. Herodotus gives a description of it:

I have dwelt longer upon the history of the Samians than I should otherwise have done, because they are responsible for three of the greatest building and engineering feats in the Greek world; the first is a tunnel nearly a mile long, eight feet wide and eight feet high, driven clean through the base of a hill nine hundred feet in height. The whole length of it carries a second cutting thirty feet deep and three broad, along which water from an abundant source is led through pipes into the town. This was the work of a Megarian named Eupalinus, son of Naustrophus.

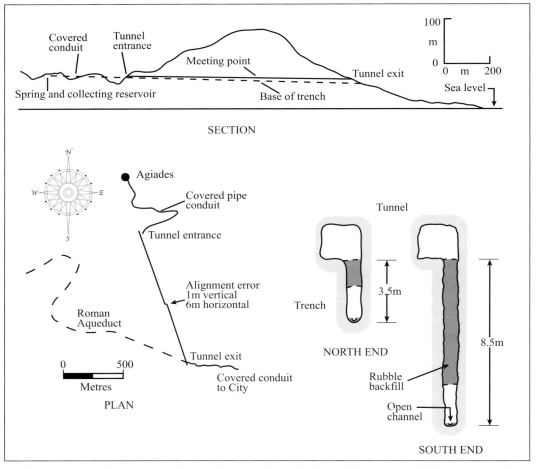

Well known to Herodotus, the tunnel of Eupalinus, on the Greek island of Samos, carried below Mt Kastro the aqueduct supplying water to the capital city. Started at each end, the tunnel excavations failed to meet by 6m horizontally and 1m vertically, requiring a bend to connect the two. The water channel is situated in the base of a trench excavated below the floor of the tunnel and deepening from north to south to give a suitable gradient for the flow of water.

Policrates' capital city Samos, on the island of Samos, occupied a site on the southern slopes of Mount Kastro overlooking the narrow, mile-wide, strip of sea separating the island from Ionia (now Turkey). A copious water supply, which Policrates wished to exploit, existed in the form of a spring to the north-west of the city, on the far side of Mount Kastro. Although the water could have been conveyed in an open channel, by following a falling contour around the mountain, this would have exposed the water supply for the fortified town to possible enemy action.

Eupalinus excavated the 1km-long tunnel, approximately 1.8m square in section, through the solid limestone rock simultaneously from both ends; he failed to meet in the middle by 6m horizontally and 1m vertically, resulting in a kink in the alignment, but still an impressive piece of surveying. The tunnel itself is almost horizontal, with a slight sag in the middle; but the terracotta channels to conduct the water were laid in the bottom of a narrow trench, running along the east side of the tunnel, excavated to a depth of some 3.5m below the floor at the northern – entry – end, and increasing to about 8.5m depth at the southern – exit – end, that gave a slope of about 0.4 per cent. The terracotta channel, in effect, occupied a second tunnel, as much of the excavated rubble was tipped back into the trench, partly back-filling it along some of its length, but leaving a space roofed with stone slabs at the bottom. The channel may have been laid in the bottom of the trench in order to control its gradient better.

A triangular-shaped reservoir, with a roof supported on fifteen square limestone pillars, collected the water from the spring and delivered it into clay pipes running along the floor of an excavated channel 1.7m deep and 0.5m wide, covered with stone slabs, which led the water by a sinuous route, 850m long and passing under three small streams, to the mouth of the tunnel, where it entered through the east wall. The reservoir may also have served as a settling tank for particles of sediment accompanying the water issuing from the spring, as the mouth of the exit channel had a higher elevation than the floor of the reservoir. Another covered underground conduit exited from the east wall, at the southern end of the tunnel, and delivered the water into the fortified city, where it may have branched out to conduct the water to a number of chosen locations.

The greatest water supply scheme in ancient times at its peak may have delivered nearly 1 million m³ per day into Rome, equivalent to almost 1,000 litres per head

Driven from both ends, the Eupalinus tunnel excavations failed to meet by 6m in the middle, requiring the deviation shown here, looking south. Also to be seen is the trench, at the base of which the water flowed in an open terracotta channel.

The Pont du Gard, some 800m in length, is a short section of an aqueduct with a total length of 40km, mostly underground, which conveyed water from a spring near Uzes to the Roman city of Nîmes in the south of France. Its three tiers of arches rise to a maximum height of over 47m.

per day, a consumption comparable with that of major cities today, including industrial usage. It was not only in Rome, but also throughout the empire that the Romans built major water supply schemes. The multiple-arched aqueduct structure dominating the centre of Segovia in Spain, nearly 30m high, continued to convey water until relatively recent times. Even more spectacular is the three-tiered structure at Pont du Gard, part of the aqueduct that supplied water to Nîmes in the south of France. In some cases, instead of opting for one of these spectacular structures to cross a deep valley, an inverted siphon of lead piping was preferred, capable of carrying water at pressures of up to 10 atmospheres or higher.

Knowledge of the aqueducts and water supply system for Rome comes from two main sources: the remains of aqueducts and reticulation systems and the treatises by the Roman authors Vitruvius and Frontinus, writing in the first century BC and the first century AD respectively. In a distinguished civil and military career, Frontinus numbered among his postings the Governorship of Britain and Manager of Aqueducts in Rome. These writings comprise valuable sources, despite the fact that some of the statements made are clearly incorrect, which in part may be attributed to difficulties in translation. For example, Vitruvius draws attention to the possibility of pressure conduits rupturing at bends, but wrongly attributes this to 'currents of air' generated in the conduit. It is possible to conclude from part of Frontinus' text that he considered the volumetric flow rate in a channel to depend only on the cross-sectional area of flow. It is highly unlikely that he believed this. It may have been a rule-of-thumb calculation, sufficiently accurate in most cases because of the degree of uniformity in the slope and roughness of Rome's conduits. It is clear from other parts of his text that he was fully aware that steeper gradients produced higher water velocities and discharges, and that channel roughness also influenced the flow. As *curator aquarum* he had the responsibility for keeping check on the discharge from the various channels leading into Rome.

The first aqueduct into Rome was the Aqua Appia built in 312 BC by the blind engineer Appius Claudius, well known for the construction of the Via Appia. Some 17km in length, the Aqua Appia brought water from Anio and the hills south of Rome. The last constructed was the Aqua Alexandrina, built by Alexander Severus in AD 226.

Although the arched structures are the most spectacular features of the Roman aqueducts, they only formed short sections of the total lengths to cross valleys in preference to an inverted siphon or to avoid a lengthy diversion around the valley, or, as in the case of Rome itself, to bring the water into the city at a high level, to a supply reservoir, giving an adequate head to ensure water flow through the city's reticulation system. The lengths of the eleven aqueducts into Rome ranged from 17km (Aqua Appia) to 90km (Aqua Marcia), totalling about 500km, with only about 30km on structure. Most of the remaining lengths consisted of open channels cut into rock or vaulted masonry conduits just below the ground surface in soils, following a gently falling contour from the source to the edge of the Campagna plains some 16km from Rome. Water tapped at the source from springs and streams was fed into collecting basins, formed if necessary by the construction of low dams, from which controlled amounts of water flowed into the contour channels. The water descended through the hills until it reached the level of the arched structure, which then conveyed it above the plains of the Campagna into the city.

In some cases two or three aqueduct channels, stacked on top of each other, shared the structure. The Aqua Claudia and the Anio Novus shared a structure, as did the Aquas Marcia, Tepula and Julia. According to Frontinus, the levels at which the channels flowed into Rome ranged from 85m above the level of the Tiber wharves for Aqua Appia to 48m for Aqua Claudia and Anio Novus (two aqueducts, Aqua Traiana and Aqua Alexandrina, were completed after the death of Frontinus: Traiana in AD 117 and Alexandrina in AD 226). Some of the aqueducts carried prodigious amounts of water: the Aqua Claudia alone conveyed over 180 million litres per day from the Sabine Hills into Rome, with similar amounts carried by Aqua Marcia and Anio Novus.

The water flowing into Rome discharged into a water tower from which it flowed into three interconnected water boxes. One box supplied the city's baths, another supplied private users – domestic and industrial – and the third, the most important, supplied water for public uses, including cisterns, fountains, military barracks and public buildings. The public could draw from the cisterns and fountains without cost, but a charge was levied on private tappings from the mains. Such tappings required the emperor's permission, but Frontinus leaves the reader in no doubt that illegal tapping was quite common, achieved by the bribing of amenable officials.

Pipes for conveying water were made of lead or baked clay. Lead pipes, ranging in width from 25mm to 230mm, were made from lead sheeting bent around a cylinder and soldered along the longitudinal joint, producing a pear-like shape in cross-section.

The Aqua Claudia and Anio Novus, both supported on these arches, delivered a total of 374 cubic metres of water a day into Rome.

Vitruvius cites clay pipes as being cheaper than lead pipes and easier to repair. His reference to the use of clay pipes for the Venter – that is the portion of an inverted siphon carried across a valley on low structure – implies that properly fitted clay pipes could carry water under limited pressure. He also dwells at length on the fact that lead could be harmful to the human system and that water conveyed by clay pipes was much more wholesome. In fact, the water ran continuously through the supply system, which, together with scale building up on the inside of the pipes, would have reduced the possible harmful effects of the lead piping on the water.

In order that water in the aqueducts flowed at a slow enough rate to maintain control of it, but fast enough to supply adequate water and to flush the channel, channel gradients had to be kept within fairly close parameters. Vitruvius states that the channel should have a gradient not flatter than 1 in 5,000, but the aqueducts serving Rome mostly had much steeper gradients than this; the Aqua Claudia, for instance, had a gradient of about 1 in 260. It called for a considerable feat of surveying to find water at a suitable level, such that it arrived at its destination at the required level to supply the reticulation system. The principal surveying instrument, the *chorobates*, consisted of a straight edge 6m long with perpendicular legs and cross-pieces at each end. The instrument was levelled by lining up plumb-bobs, attached to the cross-pieces, against lines inscribed on the ends. The instrument also had a groove on the upper edge into which water could be poured and matched against the rims to level it, in case of wind disturbance to the plumb-bobs.

2

PLYING THE WATERS

As most early settlements grew up close to streams and rivers that ensured water for domestic purposes, animals and crops, the use of the rivers for transportation would have developed naturally. Trade and peaceful contacts between settlements probably formed the earliest motivations for river travel, but these very contacts would have led to rivalries and conflicts and, quickly enough, military usage of the rivers. Floating craft were fashioned from a variety of sources: hollowed logs, tree-bark, woven reeds and animal skins were all used. Keenly interested, as ever, Herodotus was clearly intrigued by one of the means of transportation on the Euphrates:

> I will next describe the thing which surprised me most of all in this country, after Babylon itself; I mean the boats which ply down the Euphrates to the city. These boats are circular in shape and made of hide; they build them in Armenia, where they cut ribs of osiers to make the frames and then stretch watertight skins taut on the underside for the body of the craft; they are not fined-off or tapered in any way at bow or stern, but quite round like a shield. The men fill them with straw, put the cargo on board – mostly wine in palm-wood casks – and let the current take them downstream. They are controlled by two men; each has a paddle which he works standing up, one in front drawing his paddle towards him, the other behind giving it a backward thrust. The boats vary a great deal in size; some are very big, the biggest of all having a capacity of some 130 tons. Every boat carries a live donkey – the larger ones several – and when they reach Babylon and the cargos have been offered for sale, the boats are broken up, the frames and straw sold and the hides loaded on the donkey's backs for the return journey overland to Armenia. It is quite impossible to paddle the boats upstream because of the strength of the current, and that is why they are constructed of hide instead of wood. Back in Armenia with their donkeys, the men build another lot of boats to the same design.

Herodotus does not say if he himself used this means of transport, but it can confidently be assumed that he did avail himself of the Euphrates, using some means of transport. He certainly sailed great distances on the Nile. He claimed he went as far as Elephantine – a total distance of 1,500km – keenly observing the country through which the river passed, noting, for example, that from the coast inland as far as Heliopolis (i.e. the Delta) 'the country is broad and flat, with much swamp and mud, southward of which the country narrows. It is confined on one side (east side) by the Arabian mountains . . . In these mountains are the quarries where the stone was cut for the pyramids at Memphis . . . On the Libyan side (west side) of Egypt there is another range of hills where the pyramids stand; these hills are rocky and covered with sand . . .'. The journey from Heliopolis to Thebes took nine days, covering a distance of 4,860 stades (880km). The first four days passed through a level plain not more than 200 furlongs apart at its narrowest, beyond which it broadened out again.

The great rivers not only furnished a means of transportation themselves, but also provided the water for canals, which greatly widened the areas that could be accessed by water-borne traffic. The great city states of Sumer, Babylonia and Assyria owed their existence to the Euphrates and Tigris, because these rivers provided not only their water needs for irrigation and personal usage, but also the means of transport by which they could readily trade with each other and the outside world, because these rivers flowed into the Persian Gulf. Woolley estimated that the population of Ur exceeded 250,000 and may even have been twice that – far too many people to make a living solely from agriculture. Tablets found by Woolley prove Ur to have been a centre for trading and manufacture, with contacts far and wide. The factories of Ur processed raw materials, some from

Baked clay tablet found at Nippur in Mesopotamia, dating from about 1500 BC. Moated walls surrounding the town are clearly depicted, the wall following the curve of the Euphrates River on the left-hand side of the tablet. The main mid-city canal runs from top to bottom of the tablet and the square area to its right is the Mountain House temple. Part of the Nunbirdu Canal is shown at the top of the tablet.

overseas, into consumer items. A bill of lading from a merchant ship, which sailed up the Persian Gulf into the mouth of the Euphrates, and entered the city through its system of canals, showed it to have discharged onto the wharves a cargo of gold, copper ore, hard woods, ivory, pearls and precious stones.

When Hammurabi set out his famous Code of Laws he listed in the Prologue twenty-four great city states of Mesopotamia, headed by Nippur, which, because of its close association with Enlil the chief god of the Sumerians, was the paramount religious centre in Mesopotamia. Excavations at the site have shown it to have had a continual history from around 3000 BC until AD 226, but all that remains, deserted by the Euphrates, are sand-covered ruins, the river having long since carved itself a new channel towards the Persian Gulf. Perhaps the most remarkable find at this site was the earliest known map, inscribed with a reed stylus on a clay tablet. It shows with great clarity the Euphrates flowing alongside the walls of the city, and the moats and canals fed by the river, giving access by water-borne traffic into the heart of Nippur. It was in one of the two main canals, the Nunbirdu Canal, shown on the map that, according to *The Myth of Enlil and Ninlil*, the young goddess Ninlil bathed unclothed, against the advice of her mother, the wise old goddess Ninshebargunu. She warned her daughter, perhaps remembering an experience from her own past, that the sight of such beauty might incite Enlil, the father of the gods, to ravish her. Ninlil ignored the warning (otherwise there would be no story) and Enlil violated her, leaving her pregnant with the Moon-god Sin. Apparently Ninlil did not find the experience too unpleasant; further couplings between the two led to the births of more gods.

Babylon, at least for a brief time the greatest ancient city of them all in southern Mesopotamia, depended for its existence on the Euphrates and its derivative canals. The river did not merely skirt the city, but ran through its very centre, supplying the water for the moats surrounding the city and for a canal system allowing access by water-borne traffic to many locations within the city.

Canals were also cut for military purposes. The Sixth Dynasty (2323–2150 BC) Egyptian, Pharaoh Pepi I, sent an expedition up the Nile to Nubia, both to subdue the tribesmen and to bring back gold and incense. Where the Nile flowed through the first cataract near Aswan, Pepi ordered Uni, the Governor of Upper Egypt, to have canals excavated through the hard granite to overcome this obstacle. Altogether, Uni had five canals excavated, all within one year, according to the inscription he had written on the walls above the canals. The inscription still survives: 'His majesty sent me to dig five canals in the South and to make three cargo boats and four tow-boats of acacia wood of Wawat. Then the dark-skinned chieftains of Yarrat, Wawat, Yam and Mazoi drew timber for them, and I did the whole in a single year.'

Some 600 years later, around 1700 BC, an alternative solution of constructing a 3km-long slipway allowed vessels to be hauled past the second cataract near Wadi Halfa. Canals were also built linking the Nile to the important temples such as Luxor and Karnak, and the mortuary temple of Rameses III at Medinet Habu, opposite Luxor, reflecting the importance of boat processions during religious festivals.

With the breadth of their empire, and the need to transport large quantities of items such as trade goods, building materials, garrison supplies and tribute, the Romans made surprisingly little use of canals. This, despite the fact that larger loads could be transported more easily by water than over land – one horse can pull 2 tonnes on a good road free of excessive inclines, but can pull 50 tonnes by barge along a canal. Canal transportation is slow, but so is road transportation by ox-cart, oxen being the most commonly used work animal at that time for hauling heavily laden carts. Using evidence from the Price Edict of Diocletian (AD 301), it has been estimated that a journey of 500km with a wagonload of wheat doubled its price. In AD 362 Antioch experienced a famine, with grain available at a distance of only 90km.

One canal used by the Romans for navigation pursued a south-westerly route from Rome to Terracino, some 25km in length, closely paralleling the Via Appia and crossing the Pontine Marshes. Its original and primary purpose may have been to drain this area; its subsequent use for transportation no doubt owed much to some enterprising citizens realising its potential in this respect and quickly providing services for passengers and cargo. Horace, in his *Satires*, gives an amusing and lively account of travel on this canal:

Leaving the big city, I found lodgings at Aricia in a smallish pub. With me was Heliodorus, the professor of rhetoric, the greatest scholar in the land of Greece. From there to Forum Appi, crammed with barges and stingy landlords. Being lazy types we divided this stretch, though speedier travellers do it in one. The Appian is easier when taken slowly. Here I declared war on my stomach because of the water, which was quite appalling, and waited impatiently as the other travellers enjoyed their dinner.

Night was preparing to draw her shadows over the earth and to sprinkle the heavens with glimmering lights when the lads started to shout at the boatmen, who replied in kind.

'Bring her over here!'

'How many hundred are you going to pack in?'

'Whoah, that's enough!'

While the fares are collected and the mule harnessed, a whole hour goes by. The blasted mosquitoes and the marsh frogs make sleep impossible. The boatman, who has a skinful of sour wine, sings of his distant loved one, and a traveller tries to outdo him. At last the weary traveller begins to nod. The lazy boatman allows the mule to graze; he ties the rope to a stone and lies on his back snoring. When day dawns we find the barge is making no progress. This is remedied when a furious passenger jumps ashore, seizes a branch of willow, and wallops the mule and the boatman on the head and back.

It was almost ten before we landed. We washed our hands and faces in Feronia's holy spring.

A regular traveller on this route would obviously have taken the Appian Way, even if reduced to riding a mule.

A canal formed part of the port works at Ostia, but little information is available on this artificial cut. It may have extended to Rome or simply connected the port to the Tiber at some point along its length. River traffic on the Tiber at this time must have been very congested and the canal may have been constructed to relieve this.

In Roman times, the Cambridge fens and surrounding areas in England were rich farming areas, as they are today, and some of the drainage lodes still in existence are thought to have Roman origins. Of particular interest is Car Dyke, cut in the first century AD, which may have been primarily for navigation rather than drainage. Commencing from the River Cam, just north of Cambridge, it connected to Peterborough through a series of cuts and natural waterways and, from Peterborough, the Lincolnshire Car Dyke connected to the Witham, near Lincoln. Another artificial cut, the Fosse Dyke, connected the Witham to the Trent at Torksey. It was possible to navigate, by way of the Trent, Humber and Ouse Rivers, to the major Roman military garrison at York. Similar cross-section profiles for the Car and Fosse Dykes are compelling evidence that they formed part of the same navigation system. Excavations at Car Dyke have suggested a width, at water level, of 14m and a depth of a little over 2m, compared with corresponding measurements for Fosse Dyke of 15m and 2.5m. Barges moving northward on this navigation system would have been laden with farm produce for the military establishments at Lincoln, York and beyond – primarily wheat, hides and wool – and on the return journey with building materials and perhaps coal.

The Chinese have valid claims to have had the greatest hydraulic engineers in ancient times. It was the ambition of every Chinese scholar to be recorded by history as the writer of poetry and the creator of a canal. This reflected both the intangible and tangible in ancient Chinese culture – the aesthetic and the pragmatic. Canals provided the primary means of inland transportation and thus assumed as much importance to the Chinese as roads did to the Romans. The earliest sections of the Grand Canal of China were built at the time of the philosopher Confucius (551–479 BC). Upset at the moral decline of the country, as he saw it, he resigned his court position to travel to the various kingdoms, no doubt making full use of the navigable waterways, natural and otherwise, to spread his teachings, reminding adults of their ancient rites and children of the obedience and respect they owed to their parents.

In the south of the country life was relatively easy; fish was plentiful and, with copious water and a favourable climate, rice could be grown the year round. Populations burgeoned in the small kingdoms, setting up expansionary pressures, which led, inevitably, to border conflicts. The Kingdom of Wu, with its capital at Suchow, rose rapidly to power by controlling the lower Yangtze River with its war junks. Controlling the waterway brought it into conflict with its neighbours: the Chu, further up the Yangtze, the Yeh to the south and the Lu to the north. Suchow itself was situated on Tai Hu Lake, some 150km to the south of the Yangtze. King Ho Lu of Wu ordered the widening and deepening of a small creek to form a navigable waterway in 506 BC, thereby connecting Tai Hu Lake with the Yangtze River. This enabled him to move his war junks from the Yangtze down to the lake to protect his capital against his southern enemies. Notwithstanding his own considerable ambitions, Ho Lu could not have imagined that this would become the first segment of a canal eventually extending 1,700km to Beijing in the north. Ho Lu's son, Fu Chai, who was no less ambitious than his father, decided to attack the Lu to the north. At Chinkiang, where his father's canal from the south met the Yangtze, a small creek also ran to the north linking up with several lakes and the Huai River. Fu Chai had this river excavated by peasant labour to form a navigable canal 60km long, and this became the second segment of the Grand Canal.

In the fifth century BC, when the earliest sections of the Grand Canal were being constructed, China already had a well-developed east–west water transportation system based on the five major rivers: the Yellow, the Huai, the Yangtze, the Han and the West. Most ancient canal works made linkages between these rivers, creating a north–south transportation system. At much the same time that Ho Lu and Fu Chai were constructing their canals for military purposes, a start was being made in the north on the Pien canal system, which ultimately extended 430km in length and connected the Yellow River, at a point downstream of Loyang, with the headwaters of several tributaries which flowed in a south-easterly direction into the Huai River.

During the Han period (202 BC–AD 220), transportation of tribute grain increased enormously. One of the projects undertaken to cope with this, and to provide irrigation water, comprised a canal 160km long connecting the capital Chhang-an to the Yellow River. Completed in three years, it halved the transportation times from six months to three months and provided irrigation water for more than 60,000 hectares. China prospered under the Han, due in no small measure to the widespread civil engineering works put in hand to the great benefit of the country. Emperor Wu (140–87 BC) is recognised as one of the great dynamic figures in world history, and legitimate parallels can be drawn between the Han Empire and that of their exact contemporaries, the Romans. The period ended in disorder and division, which prevailed until AD 581, when a 40-year-old army general, Yang Chien, seized power, killed off all claimants to the throne and united the country.

Although the Sui Dynasty, founded by Yang Chien, survived for only a short time (AD 581–618), it accomplished remarkable achievements in canal restoration and construction.

The earliest sections of the Grand Canal in China were built for military purposes in the fourth or fifth centuries BC. Extensions and realignments over the next 1,800 years resulted in a canal 1700km in length from Hangzhou in the south to Beijing in the north, following the same route as the canal today. The world's first pound locks were built on the Grand Canal in the twelfth century.

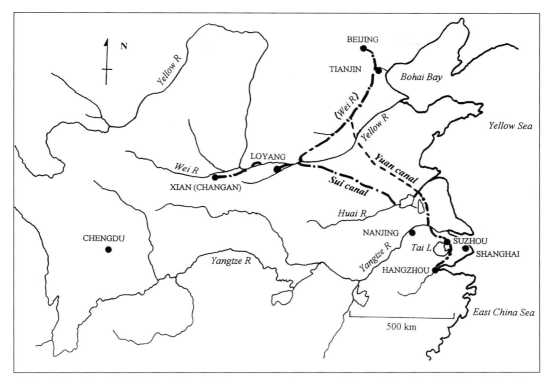

Although short lived (618–581 BC), the Sui Dynasty not only restored substantial lengths of the Grand Canal built in the Han (202 BC–AD 220) and pre-Han periods, but greatly extended it to allow uninterrupted travel by water from modern Hangzhou to Beijing. The more direct route across the North China plains, constructed in the Yuan period (1368–1260 BC), was made possible by the invention of the pound lock. (After Needham et al.)

Existing canal works were incorporated and restored, and so, by the close of the Sui period, it became possible to travel uninterruptedly by water 2,500km from Qiantang (modern Hangzhou) in the south to Tientsin (Beijing) in the north, crossing the five major rivers while doing so. As the capital Loyang was sited inland on the Yellow River, the canal assumed the shape of a giant 'Y', with the base of the stem at Loyang, one arm bearing north-east to modern Beijing and the other arm incorporating much of the older Pien Canal and Wu Dynasty canals and stretching in a south-easterly direction to modern Hangzhou. Yang Chien and his son Yang Kuang accomplished this remarkable feat by ruthlessly exploiting a labour force numbering several million – every able-bodied man over 15 years of age was forced to work on the project. Excavation equipment was primitive – mattocks, pickaxes and shovels – and the excavated earth and rock had to be carried away in shoulder-pole baskets.

A portion of the canal from Loyang to Hangzhou had willow trees flanking it on both sides, with groves surrounding the posthouses and imperial resthouses. Emperor Yang Kuang chose this section along which to make his regal voyage, to mark the opening of the canal, in a 'dragon boat' 66m long with four decks, capable of accommodating several hundred people. During the Tang and Sung Dynasties (seventh to thirteenth centuries AD), which followed the overthrow of Yang Kuang by one of his own generals, more than 165,000 tonnes of grain each year were shipped to the north from regions south of the

Yangtze. Canal works during the Tang and Sung periods were confined to maintaining the Sui system, and making minor modifications such as the erection of lock gates at river junctions. But it was during this period that the first recorded example of a lock gate, anywhere in the world, appears.

Extensive canal systems inevitably pose the problem of coping with changes in levels. The Chinese used flash-lock gates for this purpose as early as the first century BC and quite possibly before this. These consisted of simple stop-log gates – stone or timber abutments with vertical grooves into which squared logs of timber could be raised or lowered by ropes attached to their ends. Gates of this type on the Pien Canal were typically set 5km apart, obviously leading to great delays and considerable loss of water while waiting for levels to equalise (no doubt some intrepid boat-masters would not have waited for the levels to equalise). Solid wooden barriers, raised and lowered by windlass, were also used. In mountainous country, flash-lock gates imposed the need to follow a lengthy, and often circuitous, gently falling contour. A famous Chinese example was the Magic Canal, constructed in 219 BC, in Guiangxi and still in use today, although much improved and restored. Utilising a saddle through the hills, it linked the two major rivers, the Li, which flowed southwards at this point, and the Hsiang, which flowed northwards. Here they approached to within 5km. Spillways, in the form of a triangular snout built into the River Hsiang, divided its waters so that about one-third flowed into the canal, which was 5m wide and 1m deep. It had a total length, including river improvements, of 32km. The canal dropped in level from the Hsiang to the Li and, by the ninth century, eighteen flash-locks had been built to control the water flow, to be replaced by thirty-six pound locks in the tenth or eleventh century. The canal made possible, as it still does today, continuous inland navigation for 2,000km from Beijing to Canton.

The Chinese also employed the double slipway, which, in essence, consisted of a dam blocking the waterway, with gentle slopes on both approaches; boats could be hauled up the slope on one side by means of an oxen-powered capstan, then balanced precariously on the top for a few breath-catching moments before sliding down into the water on the other side. The water could be maintained at different levels on either side of the slipway, so no loss of water occurred in its use. The boats suffered severe wear and tear by being dragged up the stone ramps, and the distortions caused by this handling often badly damaged the boats or split them apart. As the goods, mostly sacks of grain, spilt out onto the ramps, they quickly found their way into the hands of organised gangs, lying in wait for such occurrences. Corrupt officials were often in their pay, and boats often experienced deliberately rough handling to cause them to shed their goods. The pressure to eliminate this problem led to the invention of the pound lock, the principle of which is to control water levels, and hence raise or lower boats, in a short section of the canal, with gates either end through which the boats can enter or leave.

The Assistant Commissioner of Transport for Huainan, Ch'iao Wei-Yo, introduced the pound lock in AD 983, at the northern end of the Pien section of the Grand Canal. A contemporary account gives a good description of the construction:

Ch'iao Wei-Yo . . . first ordered the construction of two gates as the third dam along the West River (near Huai-yin). The distance between the two gates was rather more than fifty paces, and the whole space was covered over with a great roof like a shed. The gates were 'hanging gates': (when they closed) the water accumulated like a tide until the required level was reached, and then when the time came it was allowed to flow out.

In the Yuan (Mongol) period under Kublai Khan, the introduction of the pound lock enabled the Chinese hydraulic engineer, Kuo Shou-Ching, to undertake the excavation of a new route for the Grand Canal, taking a south-easterly route across the North China plains, cutting off the arms of the great Y of the Sui canals, and halving the time taken for the 500,000 tons of tribute rice now pouring into Beijing from the south, to reach the capital. Having already completed a length of canal from Tongxian, near modern Tientsin, into the heart of Beijing, utilising both pound locks and flash-lock gates to overcome the 20m rise in level, Kuo at first followed a section of the old Sui Canal to Lingzhou, before embarking on his new course across the North China plains to rejoin the Sui route at Huaiyn, completing the work, including restoration and upgrading of the Sui sections, in 1289. The canal at the end of the Yuan period, running 1,700km from Hangzhou to Beijing, essentially followed the same route as the Grand Canal today.

The summit section of the canal presented continual problems, particularly with respect to providing adequate water for the operation of the pound locks. This may well have been a factor in the decline in the use of the canal following the Yuan period. Large grain-carrying craft increasingly used the sea routes from the south of the country to Beijing. With large craft no longer using the canal, the pound lock became disused and went out of operation, with the Chinese engineers resorting again to slipways and flash-gates. It is ironic that, as the pound lock became increasingly used in Europe to create the great waterways still in use today, it fell into disuse in China.

With their knowledge and expertise in exploiting water transportation, provided by their gift of the Nile, and their expertise in hydraulic engineering stemming from their great irrigation works, it is surely no surprise that the Egyptians should have conceived the idea of an artificial waterway to link the Nile to the Red Sea. This link would have greatly enhanced the possibilities of trade with countries to the east. To have had the level of confidence required to undertake the construction of such a waterway with such primitive resources is astounding.

According to Herodotus, the initial attempt to construct such a canal was made by Necho, a pharaoh of the Twenty-sixth Dynasty (624–525 BC), a period which saw the re-emergence of native Egyptian rule after the country had, for some 400 years, been ruled firstly by Libyans and later by Nubians and Assyrians. In a classic case of the inmates taking over the jail, the Libyans were descended from prisoners taken in warfare by the Twentieth Dynasty Egyptian Pharaoh Ramses III, who had settled the captured men in military colonies throughout Egypt. Towards the end of the Libyan period, around 750 BC, the country fragmented into a number of kingdoms and principalities, making it ripe for a takeover by the Nubians, who gradually extended their control northwards to create a united kingdom of Nubia and Egypt. But the Assyrians, commanding the most powerful and warlike army in existence at that time, had their eye on Egypt, and invasions under their kings, Esarhadon and Ashurbanipal, resulted in the destruction of Thebes and the Nubians being pushed back into their own country. By the time Necho came to power in

610 BC, Assyria itself had been annihilated by the avenging Babylonians and Medes, and Necho's father, Psammetichus I, had expelled the Assyrians from Egypt. He established what proved to be the last native Egyptian dynasty in ancient times. There followed a prosperous period during which Egypt enjoyed a flourishing overseas trade – just the right spur to take on the construction of a major canal to facilitate the movement of goods and people.

Although Necho is generally credited with being the first king to envisage the benefits, and to commence construction, of a canal to connect the Nile to the Arabian Gulf, Herodotus claimed he did not complete it. This is in contrast to the Persian King Darius, who certainly did complete it, although no doubt impressing Egyptian workers to carry out the construction under the supervision of Egyptian engineers. As Darius reigned from 522 BC to 486 BC, the canal would have been completed only about 50 years before Herodotus saw it.

In Necho's time, at a point near modern Cairo, the Nile divided into three major branches, which discharged into the Mediterranean; the eastern and western branches formed the fringes of the Delta area. About mid-way along the eastern, Pelusiac, branch, it discharged into the Mediterranean at the ancient town of Pelusium, the site of many important events in Egyptian history, including the defeat of Sennacherib's army, the defeat of the Egyptians by Cambyses and the assassination of Pompey. A freshwater lake beside the ancient town of Bubastis (now Zagazig) formed the starting point for Necho's canal, which may have terminated at the Bitter Lakes if Herodotus is correct in claiming that Necho did not complete it.

In making the claim that Necho did not finish the canal, Herodotus took a pragmatic line, mindful of the fact that the areas of his travels and where he lived for the most part, including the Greek towns along the Ionian coast, lay under Persian domination; so assigning the first completed canal to Darius must have seemed a very sensible course of action. In a later passage, while dismissing the cartographers of the time who showed a circular earth surrounded by water, with Asia and Europe being the same size, Herodotus gives his own version of the disposition of the two continents and the countries within them, and repeats the claim that the Persian king Darius connected the Arabian Gulf to the Nile by canal. The Arabian Gulf referred to by Herodotus is the Red Sea.

After calling off the construction of the canal, Necho, according to Herodotus, dispatched a fleet manned by a Phoenician crew to sail around 'Libya' (in fact Africa), starting out from the Red Sea and returning to Egypt and the Mediterranean by way of the Pillars of Hercules (Gibraltar). He claims they completed the journey in the third year of their voyage, having put to shore each autumn to sow a patch of ground, and re-embarked to resume their voyage as soon as they had harvested the grain. Thus 'Libya' was shown to be surrounded by water, confirmed to be the case, at a much later date, by the Carthaginians.

The actual length of canal may have been about half the total length of about 200km, as much of the navigation went through the Bitter Lakes, and the Red Sea may have extended considerably further north than it does today. As with all landforms, changes have occurred to the Isthmus of Suez over geological times and, even in the short period of 3,000 years, it has risen by some 3m. Further back, in Pleistocene times, a freshwater lake exited on the Isthmus, possibly fed by the waters of the Nile.

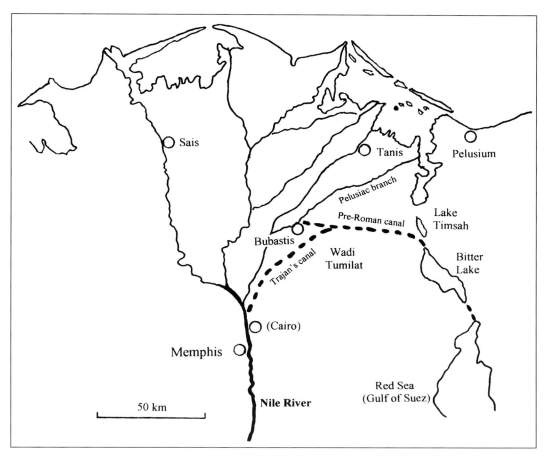

Approximate routes of the ancient canals linking the Pelusiac branch of the Nile to the Red Sea.

In order to accommodate two triremes rowed abreast, as claimed by Herodotus, the canal width at the waterline must have been about 30m, as these fearsome warships had a beam width of nearly 4m and a width between oar tips closer to 12m. The 170 oarsmen who manned these 35m-long vessels were arranged in three rows or banks on each side. The modern Suez Canal, as initially constructed, had a width at the waterline of 56m, a depth of 6m and side slopes of 1 in 2, accommodating vessels with a draught of up to 5m. Ancient vessels using the canal are unlikely to have had draughts exceeding the 1.4m of the triremes, and thus required considerably less depth of water.

Herodotus claims 120,000 Egyptians died during the excavations by Necho, presumably from disease, starvation, heat exhaustion, overwork and accident: while this may have been an exaggeration to impress readers, it should not be dismissed out of hand. In 1819 the Egyptian Viceroy Mohamed Ali had the Mahmurdieh Canal constructed, some 80km long, from Alexandria to the Nile, using forced or *corvee* labour. Something like 20,000 of the 200,000 total workforce died of sickness, starvation and overwork during the five months of the construction period. Again, during the six years from 1883 to 1889 that the French worked on the Panama Canal it is estimated that at least 20,000 died, mostly from malaria and yellow fever. More recently 100,000 prisoners are reported to have died of

Stele, now in the Ismalia museum, erected by Darius to record the building of his canal.

starvation, exhaustion and cold in the construction of the grandiose Stalinist project, the Belomor Canal, excavated in the 1930s to link the Baltic and White Seas.

When Darius entered Egypt in the service of the Persian king and Egyptian pharaoh Cambyses, he would have seen at least the remains of Necho's canal, whether it was ever completed or not. Demonstrating the astuteness that eventually gave him the Persian crown, he would have realised that such a canal opened up an important trade route to the east, in particular to India. On becoming king, he ordered its construction and, on its completion in around 500 BC, he had large granite *stelae* erected along its banks, one of which de Lesseps found in 1866 during the construction of the present Suez Canal. This stele is now housed in the museum at Ismalia, the headquarters town of the present canal. Darius makes it quite clear in the text that he caused to be written on the stele, in three cuneiform scripts on one side and Egyptian hieroglyphs on the other, both his vision and his authority in ordering the construction of the canal:

I am a Persian: with the power of Persia I grabbed Egypt. I ordered this canal to be built from the river called Pirava [the Nile[which flows in Egypt, to the sea that goes out from Persia [the Red Sea]; it was dug out according to my orders and ships sailed from Egypt through the canal to Persia, as was my pleasure.

The Egyptian text records the inauguration of the canal by twenty-four ships loaded with tribute for Persia, which arrived safely at their destination. It is likely that Darius turned this into the occasion of a state visit to Egypt by himself, and that he sailed through the canal on the leading ship. A glittering occasion, no doubt, perhaps even matching the opening of the modern canal, when twice this number of ships processed, on 17 November 1869, from the Mediterranean to Ismalia, headed by the French Imperial yacht *L'Aigle* with Empress Eugénie on board, followed in convoy by an impressive representation of European royalty. For reasons unknown Darius destroyed the final section of his canal, but it was re-opened by his son and successor Xerxes, who successfully quelled a revolt in Egypt.

The canal must have needed constant maintenance and dredging to prevent it from becoming blocked by sands blowing in from the surrounding desert, and silt settling out of the Nile waters feeding it. Over the 1,400 years of its chequered life it fell into disuse several times as the desert sands and river silts took their toll, each re-opening requiring extensive excavation. Ptolemy II (285–246 BC) had it re-opened, partly along a new course, and indeed is credited by Diodorus Siculus as the first ruler to have it completed fully to the Red Sea, previous attempts, according to his belief, having terminated at the Bitter Lakes. A port called Arsinoe (modern Suez) was built at the junction of the canal with the Red Sea and a lock constructed in the canal at this point, presumably to take up differing levels between the canal and the sea, but also perhaps to prevent sea water from entering the canal.

The Roman Emperor Trajan reopened the canal in AD 98 and had it extended to the main branch of the Nile, just upstream from Cairo. It remained in use for about a century. Finally, after the Moslem conquest, Khalif Omar granted a request from the Arab Governor of Egypt, Amru ibn el Aas, that, in order to facilitate the transport of foodstuffs to Arabian ports, he should be allowed to reopen Trajan's canal. This he did, completing the work in AD 642, the canal now known as the Canal of the Prince of the Faithful, with an excavated length of about 100km, a width of 45m and a depth of 5m. A proposal by Amru to excavate a branch canal, linking Lake Timsah directly to the Mediterranean, was forbidden by Omar, who was concerned about opening up Arabian ports to Christian ships from Europe. In AD 776 Khalif Abu Jafar Abdullah el Mansur ordered the canal to be filled at its junction with the Bitter Lakes to prevent foodstuffs, on which they were entirely dependent, reaching the insurgents of Medina. After this, it remained closed and

Never shy of self-publicity, Darius would have decreed a great ceremony to mark the opening of his canal – a spectacle repeated some 2,350 years later at the opening of the modern Suez Canal in 1869. The flotilla of ships was led through by the French Imperial yacht L'Aigle *with Empress Eugénie on board.*

gradually became subsumed by the desert. Napoleon's surveyors found traces of the canal during the French occupation at the end of the eighteenth century, as did Ferdinand de Lesseps when making his surveys for the present Suez Canal, opened in 1869.

The existence of the canal seems to have been common knowledge in classical times. Several Greek and Roman writers, other than Herodotus, made mention of it: notably Strabo, Diodorus Siculus and Pliny; all three claimed that early attempts at its construction ceased for fear that the Red Sea was higher than Egypt (by 3 cubits, according to Pliny) and that the canal would consequently cause the inundation of the land by sea water. Interestingly, Napoleon, to whom the idea of a canal appealed strongly because of the advancement of French commercial interests, dropped any thought of proceeding with its construction on the basis of a bungled survey by his engineers in 1799, which concluded incorrectly that the waters of the Red Sea at high tide rose 10m above the level of the Mediterranean at low tide.

There can be little doubt that the first ambition of Darius' son Xerxes, on succeeding to the Persian throne in 486 BC, would have been to carry on with his father's grand plan to subdue Greece and her allies and so avenge the setback at Marathon in 490 BC. Before doing so, however, he had to deal with a revolt in Egypt, which his father had been attempting to crush when he died. He also smashed a revolt in Babylon following the murder of the Babylonian satrap (viceroy), savagely destroying the city's temples, razing its walls and melting down the statue of the god Bel. With these distractions, it wasn't until 480 BC that he was able to assemble his immense army of many different nationalities (listed in great detail by Herodotus) in readiness to cross the Hellespont, together with a fleet to hug the coast of Greece and give military and logistic support to the army. Mindful of the fact that his father had lost a substantial part of a fleet (up to 300 ships and 20,000 men, according to Herodotus) attempting to round the Mount Athos peninsula in a violent northerly gale, Xerxes decided to cut a canal through the peninsula where it joined the mainland; thus, as pointed out by Herodotus, he turned the inhabitants into islanders. Herodotus describes the excavation of the canal in some detail, which suggests that he obtained a direct eyewitness account, as the work took place around the time of his birth:

In view of the previous disaster to the fleet off Mount Athos, preparations had been going on in that area for the past three years. A fleet of triremes lay at Elaeus in the Chersonese and, from this base, men of the various nations of which the army was composed were sent over in shifts to Athos, where they were put to work cutting a canal under the lash. The natives of Athos also took part.

Mount Athos is a high and famous mountain running out into the sea. People live on it, and where the high land ends on the landward side it forms a sort of isthmus, with a neck about a mile and a half wide, all of which is level ground or low hillocks . . .

I will now describe how the canal was cut. The ground was divided into sections for the men of various nations, on a line taped across the isthmus from Sane. When the trench reached a certain depth, the labourers at the bottom carried on with the digging and passed the soil up to others above them, who stood on terraces and passed it on to another lot, still higher up, until it reached the men at the top, who carried it away and dumped it. All the nations except the Phoenicians had their work doubled by the sides

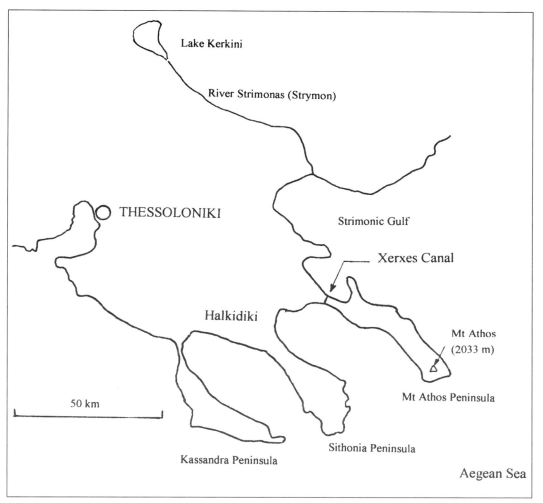

Location of Xerxes' canal across the Mount Athos Peninsula.

falling in, as they naturally would, since they made the cutting the same width at the top as it was intended to be at the bottom. But the Phoenicians, in this as in Xerxes' other works, gave a signal example of their skill. They, in the section allotted to them, took out a trench double the width prescribed for the actual finished canal, and by digging at a slope contracted it as they got further down, until at the bottom their section was the same width as the rest.

In a meadow nearby the workmen had their meeting-place and market, and grain ready ground was brought over in great quantity from Asia.

Thinking it over I cannot but conclude that it was mere ostentation that made Xerxes have the canal dug – he wanted to show his power and leave something by which to be remembered. There would have been no difficulty at all in getting the ships hauled across the isthmus on land; yet he ordered the construction of a channel for the sea broad enough for two warships to be rowed abreast.

It is a characteristic of some geological formations that they exhibit a temporary strength, allowing them to be excavated to a vertical face several metres deep, which will stand for a period of time, ranging, often unpredictably, from minutes to years, before collapsing. The marl and sandy soils at the canal site could have been readily excavated, but would have been particularly prone to collapse. It is greatly to the credit of the Phoenicians that they understood this, probably from experience, and cut the sides of the excavation back to a slope which remained stable for a period sufficiently long for Xerxes to sail his fleet through. Herodotus correctly makes the point that the other nations, in not understanding the unstable nature of this ground, greatly increased the amount of work; and it is equally likely that many of the workmen were killed, as these collapses can occur suddenly and without warning, trapping and entombing the men at the bottom of the excavation. Although sloping back the excavated faces apparently ensured they remained stable sufficiently long for Xerxes to get his ships through, they must have been too steep to remain permanently stable as the canal does not exist today. Farming of the land may also have contributed to filling in the channel. In his book *The Canal Builders*, Robert Payne states that in 1932 his father and another naval officer from HMS *Queen Elizabeth*, in hunting for signs of the canal, found 'only a tiny groove, which may never have been more than 7ft [2m] wide and 3ft [1m] deep'. They declared that they were unsurprised by this as it was a land of earthquakes, but the fact that collapses of vertical cuts were experienced during construction suggests that the sloping sides would have caved in over time irrespective of earthquakes, although these may well have hastened the process.

In asserting that the ships could have been hauled across the Mount Athos Isthmus on land, Herodotus must have had in mind the *diolkos* or ship-trackway across the Isthmus of Corinth connecting the Corinthian and Saronic Gulfs, although, surprisingly, he doesn't mention it. He would certainly have known about it, as it was constructed by the tyrant of Corinth, Periander, around 600 BC, and would have been fully operational and very actively in use during his lifetime. It is believed to have remained in use until at least the ninth century AD. Excavated remains, which can still be seen, have revealed a 4-mile (6.4km)-long paved causeway of stone blocks 3.5–5m wide with parallel grooves 1.68m apart along its length; ships were conveyed across the isthmus on cradles fitted with wheels, which ran in the grooves. Traces of double tracking indicate the provision of a passing place to facilitate two-way traffic across the isthmus.

Periander first considered constructing a 4-mile-long canal across the isthmus, but the thought of excavating through the 80m-high ridge proved too daunting, even for him, and he abandoned the idea in favour of the *diolkos*. Others who considered the possibility of constructing a canal through the Corinth Isthmus included Demetrius Poliorcetes (fourth century BC), Julius Caesar (first century BC), Caligula and Nero (first century AD) and Herodes Atticus (second century AD). Julius Caesar had plans drawn up for a canal, in the belief that it would provide him with a safeguard against flanking attacks by the Dacians during his intended march against Persia. He was assassinated before any start could be made. Although Caligula sent one of his senior officials to take accurate measurements of the isthmus, the canal became simply one of many grandiose and harebrained schemes dreamed up by him and never put into effect. He may also have been put off the project by his engineers' reports of a difference in sea level between the eastern and western ends,

The stone-paved diolkos constructed around 600 BC allowed ships to be hauled across the Isthmus of Corinth on cradles fitted with wheels.

which may have been a ploy by his engineers to discourage him from pursuing what they saw as an impossible project, or it may have been a genuine concern; in fact the sea at the western end of the present canal can exceed the eastern end by as much as half a metre, leading to a permanent west to east current through it.

Nero may have been provided with the same information, but if so it didn't deter him, which may be thought surprising in view of the fact that his interests lay primarily within the performing arts: he fancied himself as a sculptor, poet, musician, singer and tragic actor, and while his own assessment of his talents may have been a little excessive, he was not without skills in some of these areas. Suetonius, writing fifty years after the emperor's death, mentions seeing scrolls bearing verses written in Nero's own hand. In fact this much-derided emperor was a man of surprises. He murdered his stepbrother and his own mother, but ordered that gladiatorial contestants should not fight to the death. His subjects, or some of them close to him at least, benefited from enlightened, although sometimes extravagant, policies that he introduced early in his reign, perhaps due in large part to the influence of Seneca. The later emperor, Trajan, described the first five years of Nero's rule (AD 54–59) as the happiest period in the history of Imperial Rome.

Believing that the Greeks would appreciate his theatrical and artistic leanings much more than did his own countrymen, the emperor, his vanity now having reached

The modern canal through the Isthmus of Corinth, 78m high, excavated from solid rock.

pathological proportions, set out for Greece in AD 66 to tour the country and give unto them the benefit of his skills. He attended games and entered competitions, making sure the judges received proper rewards, and in return for the many prizes he won, he decided to construct a canal through the Isthmus of Corinth for the Greeks.

Nero sent in his engineers and geologists to study the lie of the land, and imported into Corinth prisoners from many parts of his empire, including 6,000 Jews captured by Vespasian, who were to be given the heaviest work, by order of the emperor. Presented with a golden spade by the local governor, and after singing suitable odes to the sea gods, Nero dug the first few spadefuls of earth, despite the fact that this initially resulted in the ground exuding a strongly red blood-like liquid, which must have caused him some alarm and certainly did so with the watching important personages, who cried out in anguish and horror. Work proceeded on the excavation of a canal up to 45m in width; but after three or four months, by which time about 500,000m³ of the total 13.5 million m³ of earth had been removed, Nero, now back in Rome, ordered the cessation of the project. His own demise followed soon after, by his own hand, muttering as he stabbed himself, 'What an artist dies in me!' He had alienated most of Rome: the army by his pacifism, the Senate by the murders of some of its members and everyone else by his love affair with the Greeks and their culture.

In 1881 a French company, after studying a number of routes, plumped for that chosen by Nero; but after toiling on it for eleven years with vastly better equipment than that available to the Roman engineers, they gave up, leaving it to a Greek company to complete. The maximum cut depth is 87m, the average nearly 60m, and the bottom width 23m. Ships up to 10,000 tons can use it and so reduce the distance between Piraeus and the Adriatic by 200 miles.

The Romans contemplated other major canal works linking to the sea, but did not carry them out. In a letter to Emperor Trajan, Pliny, then Special Commissioner for the province of Bythnia on the Black Sea, proposed excavating a canal from a nearby lake to the sea. It would have experienced a fall of some 30m over its length of 18km. Pliny envisaged transport on the canal of such heavy items as marble, farm produce, and timber for building, thus obviating the great difficulty and expense of the road haulage then being used. Trajan expressed the concern that the canal might drain the lake to the sea and offered the services of an engineer to review the project. Pliny accepted the offer, but in the meantime modified his proposed scheme by taking the canal to the edge of the nearby river, leaving an earthen barrier across which the goods would be transferred by hand from the canal boats to the river boats and thence taken down to the sea. Pliny died shortly after and the proposal died with him.

Tacitus mentions an even more ambitious scheme, which contemplated connecting the Rhine and Saône Rivers:

> To keep the troops busy the Imperial Governor of Lower Germany, Pompeius Paulinus, finished the dam for controlling the Rhine, begun sixty-three years previously by Nero Drusus. His colleague in Upper Germany, Lucius Vetus, planned to build a Saône–Moselle canal. Goods arriving from the Mediterranean up the Rhone and Saône would thus pass via the Moselle into the Rhine, and so to the North Sea. Such a

waterway, joining the western Mediterranean to the northern seaboard, would eliminate the difficulties of land transport.

Again, it never happened – according to Tacitus because of jealousy on the part of the Imperial Governor of Gallia Belgica, through which the waterway would have passed. Another deterring factor in both this and the Bythnia scheme may have been that the Romans did not know how to overcome the large water-level changes that both schemes would have experienced. The precise intended route for the Rhone/Saône canal is not known, but the Canal d'Este, which does today connect the Saône and Moselle Rivers, rises 40m in 3km from the Moselle, before dropping 137m in 50km to the Saône. Such changes in level require the use of the pound lock, which was not introduced into Europe until the late fourteenth or early fifteenth century.

While nature generously provided the rivers and seas to be exploited for transportation, port and docking facilities still had to be constructed to enable the craft to be loaded and unloaded, and to provide maintenance and construction facilities and perhaps shelter in stormy weather. Early structures may have simply consisted of a few logs set into the bank of a river, but, after these had been washed away a few times, more sophisticated structures would have been devised, consisting of poles driven into the river bank supporting a platform of cross-timbers and decking. Where port facilities for ocean-going vessels did not have natural storm protection, this had to be provided by the construction of breakwaters, creating a harbour within which the ships could shelter. Herodotus mentions such a breakwater, which he must have known well, at Samos: 'there is the artificial harbour enclosed by a breakwater, which runs out into twenty fathoms of water and has a total length of over a quarter of a mile.' This 300m-long breakwater, built in the reign of Policrates (540–523 BC), was 35m high above the seabed, a testament to the great skills shown by Greek harbour builders in selecting and exploiting natural harbour sites. Policrates became the most powerful Greek ruler in his time, the breakwater at Samos serving his 150 warships, and giving him complete control over the Aegean. Other examples of Greek harbour construction include Kenchrae on the Saronic Gulf and Piraeus, the port for modern Athens, where port facilities existed as early as the fifth century BC.

Harbour facilities taken over by the Romans were improved and reconstructed to such an extent that very little evidence of pre-Roman port facilities remain. There are also very few descriptions in ancient writings. An exception is the description by the Jewish historian, Josephus Flavius (AD 37–95), of the harbour at Caesarea, built by order of King Herod. Underwater archaeology has confirmed Josephus' description, as well as revealing additional information. Breakwaters were formed in part by huge ashlar blocks, up to 50 tons in weight, fastened together with iron clamps, and in part by hydraulic cement, made up of lime, red soil and volcanic pumice, placed between wooden shuttering. In order to economise on construction materials, a cellular form of construction was adopted, with hollowed compartments 20m by 30m to be filled by wave-carried sand and ultimately paved over. A series of channels fitted with sluice gates ensured that only silt-free water entered the harbour and so prevented the build-up of silt on the harbour floor.

Surpassing all other works taken over by the Romans was the great city and port of Alexandria, conceived by Alexander when he visited Egypt in 331 BC, but built after his

The Pharos lighthouse at Alexandria, one of the seven wonders of the ancient world, completed in 283 BC under Ptolemy II. Although its exact appearance is conjectural, it is known from various descriptions to have consisted, as shown here, of three tiers, the lowest square in plan, the middle octagonal and the top circular, topping out at perhaps 135m height. Fuel may have been taken to the top by animal portage up an internal spiral ramp. (From Hammerton)

Harold Oakley

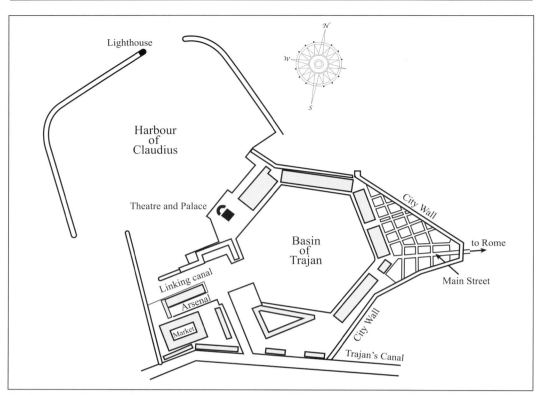

The harbour of Portus at Ostia, the port for Rome at the mouth of the Tiber, built by Claudius in AD 46. The hexagonal inner basin with warehouses and other structures was added later by Trajan (AD 98–117).

A street in Ostia lined on either side by brick-built warehouses and depots for the storage of foodstuffs. (From Hammerton)

death. A mile-long breakwater and causeway, built on a reef, protected incoming shipping from strong north-westerly winds, and provided a causeway linking the harbour to the Isle of Pharos and the famous lighthouse. One of the seven wonders of the ancient world, the lighthouse stood for almost 1,500 years until, like many masonry structures in Mediterranean countries, it crashed down in an earthquake, in this particular case in the thirteenth century AD. Influenced by the positioning of this famous edifice, the Romans frequently built lighthouses in association with their harbour works, in order to guide ships safely into port, rather than the more common modern purpose of lighthouses warning against danger.

More concerned with location than site features, the Romans did not baulk at implementing massive civil engineering works to create their harbours. In his description of the harbour of Centumcellae, part of the inner basin of the modern port of Civitavecchia, Pliny refers to huge stones transported to the site in broad-bottomed barges and sunk one upon the other to form a solid substructure, while Vitruvius describes a construction technique, for use in rough or fast-flowing waters, of casting large concrete blocks above water, on a sand platform confined by marginal timber walls. The concrete was allowed two months to set; then the confining walls were cut away and the sand ran out, helped by wave action, allowing the concrete blocks to settle onto the sea bed. One of the most important discoveries by the Romans was that naturally occurring volcanic ash, known as *pozzuolana*, had the properties of a hydraulic cement, capable of setting hard under water. Vitruvius describes the construction of jetty walls under water using *pozzuolana* concrete mixed in mortar troughs and tipped into a space within coffer-dams, created by driving oaken stakes into the seabed with ties between them, until the concrete displaced all the water.

The main port for Imperial Rome occupied a site at the mouth of the Tiber, close to Ostia. In AD 46 silting up of the old harbour at Ostia prompted the much-maligned Emperor Claudius to have two great moles, formed from riprap in 5–6m of water, thrown out into the sea to enclose an area 1,000m², and new harbour works constructed within their confines. Pliny describes how Caligula ordered a huge vessel, built specifically to bring an Egyptian obelisk to Rome, to be sunk by weighting it with *pozzuolana* and rubble and incorporated it into the harbour works. Using this as a foundation, he then placed on it square ashlar blocks, on which he had a lighthouse built, its beacon of burning pitch 60m above sea level and visible many miles out to sea.

Trajan commissioned major extensions to the harbour, notably the construction of an inner harbour, hexagonal in form and 700m broad, with spacious warehouses and other associated structures. Concrete and brick were used to build up the key walls, with projecting stones inset at intervals and pierced with horizontal holes for mooring ropes and bollards. The floor of the basin, some 4–5m below water level, was paved with stone to facilitate dredging, which must have required constant operation to keep the harbour free of the silt and sand brought down by the river. Like many ancient harbours it has now succumbed to the silt and lies two miles inland, the ruins still visible amongst the pine trees. In its day it must have been the world's busiest port, bringing in goods from all around the known world, exotic spices and silks, more prosaic sacks of corn and jars of oil or wine, blocks of stone or marble to satisfy the urge to build Rome into the greatest city the world had ever seen, and wild animals from Africa to satisfy the blood lust of its populace.

3

CROSSING THE WATERS

Early man would no doubt have made good use of fallen logs to traverse streams, and it would have been a simple enough step to place a log at a suitable crossing point where one did not already exist. Further developments would have included forming a bridge with two parallel logs laid some distance apart across the stream, with a decking of timber branches spanning them, and abutments of horizontal timber logs laid along the river bank in the direction of the stream, giving firm support to the ends of the timber beams. Where the bank was soft or unstable the abutments would have been strengthened by securing them to timber piles driven into the riverbank. Other improvements included the provision of timber handrails to make the crossing of the bridge safer, and hand-cut planking to provide a more even flooring. Stone slabs, where available, provided an alternative to timber logs to form

Clapper bridges seen today, such as the Tarr Steps in Devon, probably date from the Middle Ages or even later, but are of a type dating back to much earlier times. Although subject to destruction by floods, they are easily re-erected.

structures known as clapper bridges, sometimes consisting of several spans. Examples of such bridges, which can be seen today, are likely to be of a more recent age as they are prone to demolition by floodwaters, but nevertheless they represent a form certainly dating back many thousands of years. They had the advantage that they could be easily reconstructed or replaced once the floodwaters had passed.

Simple suspended spans of vines or ropes have also undoubtedly been used to cross rivers and streams since the time of early man and are still constructed in some parts of the world today, particularly for conveying foot traffic across rivers confined within deep gorges. The most elementary type would have been a single cable securely tied to trees or stable rocks on each side of the river, allowing a man to cross hand over hand; or two cables, one above the other, the lower one supporting the feet and the upper one providing hand holds. A further development was the use of a pair of cables, a metre or so apart, with a woven mat or some other form of flooring such as timber saplings spanning them. A cable or cables at a suitable height above the flooring provided handrails. These basic forms, nourished by human ingenuity, sprouted numerous variant branches, many of which spanned the gorges of the eastern Tibetan massif, an area that Needham identifies as the focus of origin of ancient suspension bridges.

The Chinese introduced a major innovation by fashioning ropes from bamboo. Strips of outer, silica-rich layers of bamboo were plaited together and wrapped around a core of the inner culm of the bamboo, tension in the cable causing the outer portion to grip the core tightly. Three or more of these ropes, each about 50mm thick, twisted together formed a cable with a tensile strength greatly in excess of hemp rope and with a silica-rich surface highly resistant to erosion. Needham cites the famous example of the 3m-wide bamboo suspension bridge at Kuanhsien, with a total length of 320m and a maximum single span of no less than 60m. Its flooring is supported by ten bamboo cables, each 165mm in diameter, and it has handrails on each side consisting of five similar ropes. It is supported, some 15m above river level, by timber trestle piers and one pier of granite masonry. Needham puts its age as certainly pre-Sung and possibly as early as the third century BC, although it has been subject to routine annual maintenance and extensive renovation from time to time throughout history.

A more robust material – wrought iron – was used in China for forming suspension chains as early as the sixth century AD and possibly as early as the first century AD, in the Han period. Either way this makes it almost certainly the first, and perhaps the only, use of iron for structural elements in the ancient world. Needham lists an impressive number of such bridges, particularly in the Yunnan and Szechuan provinces, built from the time of the Sui Dynasty in early sixth century AD onwards. Spans ranged up to 100m. Two types of cable are reported: iron-link chains and pin-connected iron rods. Massive stone abutments provided anchorage for the chains at each end of the bridge; in some cases, where a bridge spanned a deep rocky gorge, holes drilled into the rock secured the cables.

Unlike modern suspension bridges, the floor or deck of these ancient bridges rested directly on the cables and thus followed the same catenary curve. The place and time of origin of the horizontal deck suspended by hangers from the cables is obscure, but the first known example of a major bridge of this type was constructed in the early nineteenth century across the Merrimac River in Massachusetts, with a span of 74m.

Where urgency dictated construction, such as in military campaigns, or where great depth of water or poor riverbed conditions precluded the possibility of founding piers in the water, one answer to bridge construction, which seems to have been quite common in the ancient world, was the floating or pontoon bridge. The Hellespont represents the most celebrated of these. Herodotus gives a vivid account of its construction:

He [Xerxes] prepared to move forward [from Sardis] to Abydos, where a bridge had already been constructed across the Hellespont from Asia to Europe. Between Sestos and Madytus in the Chersonese there is a rocky headland running out into the water opposite Abydos. It was here not long afterwards that the Greeks under Xanthippus the son of Ariphron took Artayctes the Persian governor of Sestos, and nailed him alive to a plank – he was the man who collected women in the temple of Protesilaus at Elaeus and committed various acts of sacrilege. This headland was the point to which Xerxes' engineers carried their two bridges from Abydos – a distance of seven furlongs. The Phoenicians constructed one bridge using flax cables, while the Egyptians raised the other with papyrus cables. The work was successfully completed, but a subsequent storm of great violence smashed it up and carried everything away. Xerxes was very angry when he learnt of the disaster, and gave orders that the Hellespont should receive 300 lashes and have a pair of fetters thrown into it. I have heard before now that he also sent people to brand it with hot irons. He certainly instructed the men with the whips to utter, as they wielded them, the barbarous and presumptive words: 'You salt and bitter stream, your master lays this punishment upon you for injuring him, who never injured you. But Xerxes the King will cross you, with or without your permission. No man sacrifices to you, and you deserve neglect by your acid and muddy waters.' In addition to punishing the Hellespont Xerxes gave orders that the men responsible for building the bridges should have their heads cut off. The men who received these invidious orders duly carried them out, and other engineers completed the work. The method employed was as follows: galleys and triremes were lashed together to support the bridges – 360 vessels for the one on the Black Sea side, and 314 for the other. They were moored slantwise to the Black Sea and at right angles to the Hellespont, in order to lessen the strain on the cables. Specially heavy cables were laid out both upstream and downstream – those to the eastward to hold the vessels against winds blowing down the straits from the direction of the Black Sea, those on the other side, to the westward and towards the Aegean, to take the strain when it blew from the west and south. Gaps were left in three places to allow any boats that might wish to do so to pass in or out of the Black Sea.

Once the vessels were in position, the cables were hauled taut by wooden winches ashore. This time the two sorts of cable were not used separately for each bridge but both bridges had two flax cables and four papyrus ones. The flax and papyrus cables were of the same thickness and quality, but the flax was the heavier – half a fathom of it weighed 114lb. The next operation was to cut planks equal in length to the width of the floats, lay them edge to edge over the taut cables, and then bind them together on their upper surface. With that done, brushwork was put on top and spread evenly, and a layer of soil was trodden in hard all over it. Finally a paling was constructed along each side, high enough to prevent horses and mules from seeing over and taking fright at the water.

Xerxes' army took seven days and nights to cross the bridge.

In resolving to cross the Hellespont in this way, Xerxes had before him the example of his father who, some thirty years before, had achieved the hardly less impressive feat of crossing the Bosphorus with his army, on a floating bridge of boats designed by the Samian Mandrocles, to invade Macedonia and Thrace. The account of this crossing, given by Herodotus, is briefer than that for the Hellespont, probably because he had fewer details of this earlier crossing and also it was of less interest to the Greeks. Unfortunately, a picture commissioned by Mandrocles showing the Great King surveying his army crossing the Bosphorus has not survived.

Darius (reigned 522–486 BC) inherited a great empire, but a turbulent one, conquered by his predecessors. The first Great King Cyrus (546–529 BC), having first overcome Croesus and annexed Lydia, went on to subjugate Babylonia and Ionia, and eventually the whole of Asia Minor, to which his son Cambyses (529–522 BC) added Egypt. After overcoming a palace revolution against Cambyses, who died returning from Egypt to put it down, Darius had to direct his attentions firstly to quelling a number of revolts in order to secure the empire and then to setting up an effective organisational structure. With these matters in hand he could contemplate extending his empire into Europe. To do this he had to cross the Bosphorus. He did so in 512 BC.

His bridge-building operations on this campaign did not stop at the Bosphorus. Before crossing over into Europe Darius ordered a fleet of his ships to sail into the Black Sea, to the mouth of the Danube, and to sail up river to a point where the main stream divided.

Although Xerxes' crossing of the Hellespont by pontoon bridge is the more celebrated, he was following in the footsteps of his father, Darius, who, some thirty years earlier, had crossed the Bosphorus, seen here, by the same means.

There he ordered his men to construct a bridge (most likely another pontoon bridge) across the river and await his arrival. Darius arrived at the bridge after overcoming resistance from a Thracian tribe named by Herodotus as the Getae, who were immediately enslaved and forced to accompany his army on its march. After crossing over the bridge into Scythia, Darius ordered its destruction, but one of his commanders, a man called Coes, bravely suggested that they should leave the bridge intact, guarded by the men who built it. His suggestion was based not on the expectation of defeat by the Scythians or any wish on his part to remain behind to guard the bridge, but against the possibility that they would fail to find the Scythians, and endless and indeterminate marching to do so could get them into difficulty.

Coes must have been greatly relieved when Darius not only went along with his suggestion, but also promised to reward him on his return home for his wise counsel. Darius then took a leather strap, into which he had tied sixty knots, and instructed the men delegated to guard the bridge to untie one each day, and if he had not returned by the time they untied the last knot, they had his permission to sail home. In the event, the Scythians adopted a policy of retreating, keeping a day or so ahead of the enemy, blocking up the wells and springs as they passed, as well as destroying the pasture and using cavalry to harass the flanks of the Persian army, quickly withdrawing and melting away after each attack. Realising the futility of his enterprise, and concerned that the Scythians, who knew the country well, could double back and destroy the bridge, Darius decided to withdraw across the Danube, leaving a token force behind in Scythia. Emboldened, the Scythians pursued the Persians but, instead of engaging them, made directly for the bridge and, as they were mounted and the Persians mostly on foot traversing unfamiliar territory, they reached the bridge before the retreating army.

Rather than attacking the bridge they set about persuading the Ionians on the ships guarding it to demolish it and return home, pointing out to them that the sixty days had in fact expired; but, in any case, they had nothing to fear as they, the Scythians, intended to settle with Darius in such a way that this would be his last campaign. Some of the Greeks favoured this proposal, arguing that they could then go on and free Ionia. But wiser counsel prevailed and they decided to remove a portion of the bridge (about a bowshot in length, according to Herodotus) on the Scythian side of the river, both to trick the Scythians into believing they were demolishing it and to prevent the Scythians from forcing a passage over the river. Satisfied that the Ionians were indeed breaking up the bridge, the Scythians went in search of Darius' army and, perhaps by design, and almost certainly to their good fortune, failed to make contact. The Persian army eventually reached the crossing with some difficulty, but without encountering any attacks by the enemy. Arriving in the dark, to their consternation they saw only the broken portion of the bridge; but a shout from a man with a tremendous voice roused the guards on the ships, who immediately mobilised the vessels to start ferrying the army across, at the same time making good the broken section of the bridge. Darius marched back through Thrace to the coast and took ship across the Hellespont for Asia, leaving an army of 80,000 men behind in Europe under the command of a distinguished Persian general named Megabazus.

In 490 BC – twenty-two years later – Darius again took ship across the Hellespont, in a reversal of direction, to attack Athens in reprisal for their support of the Ionian revolt of

499 BC against the Persians. Darius not only put down the revolt by defeating the Ionian navy and sacking their major city, Miletus, but he also set up a system of semi-autonomous governments in Ionia, which lost no time in renouncing their close alliance with Athens. Darius landed his troops from their fleet of transports in the bay of Marathon, 32km north-west of Athens, which, in effect, initiated the Graeco-Persian wars. A force of 10,000 Athenians routed the Persian invaders, who quickly re-embarked and returned whence they came, leaving the Greeks to relay the good news to Athens by means of the stout-hearted Phidippedes, who dropped dead from exhaustion at the end of his run, but had the posthumous satisfaction of giving rise to a famous athletics event.

A tradition of pontoon bridge building existed in the Levant and Mesopotamia in ancient times, reflecting the high degree of military activity in the region and the difficulty of founding piers in the beds of large rivers such as the Euphrates and Tigris. Marching armies in Asia Minor encountered many rivers and made frequent use of floating bridges, either already in place or hastily constructed. Alexander, in his campaigns, carried boats with him, broken down into manageable pieces, to allow rapid crossing of rivers too deep to ford or where no bridge already existed. He crossed the Indus on a floating bridge built by his own engineering corps, headed by Hephaistion, his army now numbering some 80,000, with 30,000 camp followers plus pack animals and elephants.

Roman military engineers, although more celebrated for their magnificent masonry arches, also knew of the possibilities of pontoon bridge construction as a quick method of providing a river crossing. So, it seems, did their enemies. In his Gallic War commentaries Caesar describes his campaign against the Helvetii, a Celtic people who had been driven from their homes in Germany into Switzerland and had then decided to migrate westward to settle near the Atlantic coast. Despite charges by his critics of unnecessary and unprovoked aggression, Caesar decided his military prestige depended upon stopping them. After slaughtering tens of thousands of them, he drove the remainder back into Switzerland. When the Helvetii reached the Saône River, which Caesar describes as sluggish *beyond belief*, they proceeded to cross by tying rafts and boats together This pontoon bridge took twenty days to construct. By the time Caesar learnt of this, three-quarters of the Helvetii had succeeded in crossing the river, the remaining quarter quickly perishing at the hands of the Romans and their allies. In order to pursue those who had crossed, Caesar claims to have bridged the river in one day, which probably means he simply repaired or strengthened the existing Helvetian structure. Alarmed by this, the Helvetii sent deputies to him, but they received little comfort and Caesar continued his pursuit of them.

The procedure adopted for building floating bridges is described by Dio Cassius:

The channels of rivers are very easily bridged by the Romans since this is always practised by the soldiers on the Danube, the Rhine and the Euphrates just like any other military exercise. The manner of construction, which is not familiar to everyone, is as follows. The boats, by which the river is to be bridged, are flat-bottomed and are anchored a little way upstream, above the spot where the bridge will be built. When the signal is given, they release one boat to drift in the current near the bank they are holding. When it has floated into a position opposite the place to be bridged, they throw into the stream a wicker basket filled with stones, attaching a cord so it serves as an

Floating, or pontoon, bridges were common in the ancient world, as they could be constructed rapidly for military purposes and they avoided the need to construct piers in flowing water. This nineteenth-century engraving depicts such a bridge crossing the Indus River.

anchor. The boat, secured in this way, stays near the bank and a base is paved immediately up to the landing with the boards and materials for the bridge, which the boat carries in abundance. Then they release another boat at a little distance from the first, and another from that one, until they have driven the bridge to the opposite bank.

On a more frivolous note, Caligula, not to be outdone by Xerxes' feat in crossing the Hellespont over 500 years earlier – a feat which still captured people's imagination – had a bridge of boats thrown across the Bay of Naples for the sole purpose of staging parades, at the heads of which he strutted in fancy dress.

There exists relatively little evidence of fixed bridges in pre-Roman times, which is not surprising as these would have been mostly of timber and long since rotted or, in many cases, did not survive long before being swept away by floodwaters. The earliest detailed description of an all-timber bridge was a temporary military structure thrown across the Rhine in 55 BC by Julius Caesar's engineers, and described by Caesar in his *Commentaries*. Timber piles about 400mm in diameter were joined in pairs about 600mm apart and driven from rafts at raking angles upstream and downstream, and their tops, about 12m apart, were joined by stringers 600mm thick to form a trestle. Beams with their ends resting on the stringers spanned the 6–7m gap between trestles and flooring planks were laid on these beams. Cutwater piles, driven upstream and downstream of each support trestle, protected the bridge from river ice or floating logs. According to Caesar, the 500m-long bridge took ten days to construct, an impressive demonstration of organisational and construction skills which clearly contributed to Roman military successes.

One of the best-known bridges in ancient times must have been the one spanning the Euphrates, giving access between the sections of the inner city of Babylon occupying either side of the river. Standing on the bridge at times of peak traffic, one would have been assailed by a babble of languages and a mixture of people in colourful national dress from countries as far apart as Greece and India, as well as the occasional homesick Israelite. One might even have seen Herodotus stroll by, deep in discussion with a local sage, who may have imparted the information on the bridge that he gives in his writings:

Nitocris, however, when she was having the basin dug for the lake, had the foresight to make that work the means of getting rid of the inconvenience [of having to cross the river in Babylon by boat] as well as of leaving another monument of her reign. She ordered long stone blocks to be cut, and when they were ready and the excavation complete, she diverted the river into the basin; and while the basin was filling and the original bed of the stream was drying up, she built with burnt brick, on the same pattern as the wall, an embankment on each side of the river where it flowed through the city, and also along the descent to the water's edge from the gates at the ends of the side streets; then, as near as possible to the centre of the city, she built a bridge over the river with blocks of stone which she had prepared, using iron and lead to bind the blocks together. Between the piers of the bridge she had squared baulks of timber laid down for the inhabitants to cross by – but only during daylight, for every night the timber was removed to prevent people from going over in the dark and robbing each other. Finally, when the basin had been filled and the bridge finished, the river was brought back into its original bed, with the result that the basin had been made to serve the queen's purpose, and the people of the town had their bridge into the bargain.

According to Koldewey, the 115m-long bridge rested on seven piers, 9m by 20m in plan, and closely spaced to support the simple timber spans of beams and decking. A composite construction of large stone blocks and baked brick, these massive piers avoided the problems of decay and damage from floating debris, which timber trestles would have suffered, but would have introduced a major problem in restricting the river flow and consequently caused swift currents, which must have consistently threatened to undermine the foundations of the structure.

The citizens of Rome in 500 BC would have given much for the Pons Sublicius, built in 621 BC across the Tiber, or to have had the removable spans of the Babylon bridge as described by Herodotus, when Lars Porsena arrived on the opposite bank with his Etruscan army intent on besieging the city. The bridge may well have been otherwise similar in construction to its Babylonian contemporary. Came the moment; came the man:

> Then out spake brave Horatius,
> The Captain of the Gate:
> To every man upon this earth
> Death cometh soon or late.
> And how can man die better
> Than facing fearful odds,
> For the ashes of his fathers
> And the temples of his gods?

Accompanied by two volunteers, Horatius marched out to defend the bridge, while others destroyed by 'axe and lever' the timber spans behind them. In Macaulay's stirring account the two volunteer soldiers managed to dart back before the bridge fell, but Horatius wasted time apparently contemplating whether he should finish off the 'thrice thirty thousand' Etruscans on his own or plunge into the river. He sensibly chose the latter course and survived to be honoured by a grateful Roman public.

Although timber has excellent qualities as a building material, strength limitations limit simple spans to about 6m. Above this, the self-weight of the timber beam, together with the imposed load, causes it to sag or even break. As timber offers advantages in handling and speed of construction, and was sometimes the only suitable building material readily available, there must have been a constant challenge to use it for spans in excess of 6m and avoid a multiplicity of mid-stream supports. Solutions devised to overcome this problem included both timber cantilevers and arches made up of shorter timber pieces.

An excellent example of a Roman timber arch bridge is shown in relief on Trajan's column in Rome. An avid builder, Trajan employed the services of Appollodorus of Damascus for many of his works, including this bridge across the Danube, the purpose of which was to give access for his armies into what is now Romania in order to subdue the Dacians. According to Dio Cassius, writing a century after the bridge was completed, it had massive squared stone piers 45m high, 18m wide and 15m thick. In building piers of

This arched timber bridge surmounted by a timber truss parapet, and resting on stone piers, is depicted on Trajan's Column. It crossed the Danube. (From Lepper and Frere)

these dimensions the Romans clearly recognised the huge thrusts the 33.5m-long arch spans would exert on the piers. Struts, linking the three concentric timber arches of each span, ensured their combined action and also served to support the deck beams. Another interesting feature was the triangulated truss parapet, but the scale of the bridge in the relief is obviously distorted and the parapet is not likely to have had any structural role. Having built the bridge to subdue the Dacians, the Romans had to destroy it thirty years later to stop the incursions of the Dacians into the Empire.

The Roman bridges still in existence today, and there are many of them, are of masonry construction, and their very survival attests to the fact that the Romans perfected the art of the *voussoir* arch, in which the arch ring is composed of carefully cut and fitted truncated wedge-shaped blocks. With firm abutments and carefully shaped blocks, this form of arch does not need mortar between the blocks. It suffers the drawback, however, that it has to be supported during erection by an arch-shaped centring, which was usually a timber structure resting on the abutments. The arch ring is built up progressively from the two abutments until the top *voussoir* or keystone is placed, whereupon it becomes self-supporting and the centring can be removed. Despite the fact that it derives its strength solely from its own weight and from the loads imposed on it, it is a remarkably resilient structure capable of tolerating severe distortions without collapsing. Surviving *voussoir* arch rings are often to be seen standing proudly in otherwise completely ruined ancient structures. Depending on its width, a bridge span will be made up of many abutting arch rings to form, in effect, a vault. Although the Romans proved themselves the supreme masters in the use of the *voussoir* arch in bridge construction, they did not invent it. The Etruscans used it for bridge construction as early as 1000 BC and they probably passed it on to the Romans. It is thought that the Etruscans originated in the Eastern Mediterrranean, possibly taking their knowledge of the *voussoir* arch with them when they moved on to Italy.

Roman engineers achieved an unchallenged pre-eminence in the use of the true arch form despite their almost obsessive adherence to the semi-circular *voussoir* ring. As Vitruvius expressed it, bridges had 'archings composed of *voussoirs* with joints radiating to the centre'. This can describe a segment of a circle as well as the full semi-circle, but the segmental form does not appear to have been used by the Romans. It is greatly to the credit of their engineers that they created such imposing and often beautiful masonry bridges, in an immense variety of settings, while remaining faithful to the semi-circular arch with its immutable relationship between span and rise. Despite extensive use of concrete in buildings from the first century AD onwards, the Romans rarely used this material to form their bridge arches, but made good use of it in their bridge foundations.

Several bridges built by the Romans across the Tiber survive intact and are still in use today. The seven-span Pons Milvio continues to carry the heavy traffic of the Via Flaminia and, during World War II, it carried the entire military traffic of the Italian, German and Allied armies. Built in 109 BC of *tufa* and travertine, its arched spans range from 16m to 24m. Small arched openings over the piers help to pass the floodwaters and also lighten the weight on the foundations. By modern standards the roadway is narrow, with a width of only 7m between parapets. Although some restoration has been carried out on this bridge, notably in the fifteenth century, most of it is still original. As with any major bridge, it has witnessed many important events in history. In 63 BC the Patrician Cataline, impoverished and an undischarged prisoner on an embezzlement charge, was allowed, after a number of attempts, to stand for the consulship against Cicero. His programme of land distribution and general cancellation of debts ensured his defeat by the conservatives and moderates, and, unable to stomach this reverse, he plotted revolution, having mustered the support of a number of other conspirators. Five of these were caught in the city by Cicero and immediately put to death, but Cataline fled northwards across the Pons Milvio with his followers, only to be caught and destroyed by government troops stationed at both ends of the Apennines. This event is particularly noteworthy for the eloquent and persuasive speeches and writings by Cicero that contributed much to quashing the conspiracy.

A more profound event that completely changed the course of history occurred on the Pons Milvio in AD 312, with the defeat here of Maxentius by Constantine, making him undisputed emperor of the west the very day after he had seen the flaming cross in the sky, which caused him to embrace the Christian religion.

Another famous Roman bridge still standing is the very handsome Pons Fabricius or, to give it its modern name, Ponte Quattro Capi, from the four-headed god, Janus, carved on the parapet. Its original name derived more appropriately from Lucius Fabricius, city engineer to Republican Rome, and thus responsible for building it. Built in 62 BC, during the consulship of Cicero, to connect the island of Aesculapius to the left bank of the Tiber, it has an inscription indicating that it was built by a private contractor, stating that the contractor would have his deposit returned after forty years if the bridge remained intact.

Built in 109 BC across the Tiber, the Pons Milvio has witnessed many important events in history, including the defeat of Maxentius by Constantine on the day in AD 312 the latter witnessed the flaming cross in the sky, causing him to embrace Christianity.

On that basis he should have been refunded his deposit many times over. As with most of the Roman bridges, mortar was not used between *voussoir* blocks, which required the blocks to be faced very accurately to give closely mating joints; the blocks were, however, bonded laterally with iron cramps. With a depth of a *voussoir* arch ring only about one-tenth of the arch radius, this bridge represented a daring design for the time. The two main arches, spanning 24.2m and 24.5m, are supplemented at each end by smaller arches, which are now buried in the banks of the river. A relieving arch pierces the spandrel above the central pier. The heavy cutwaters are typical for Roman bridges; they are triangular and sharply pointed when facing upstream and are rounded when facing downstream.

It is not possible to leave Rome itself without mention of the Pons Aelius, perhaps the most impressive of all the bridges in the capital. It is now better known as the Ponte Sant' Angelo. Completed in AD 134 by Emperor Hadrian to connect the Campus Martius to the splendid castle he had designed himself as his own mausoleum, it crosses the Tiber with seven arches, five of which are still visible, with spans of up to 19m. Although it was largely rebuilt in the fifteenth century at the order of Pope Nicholas V, its three central spans are considered to be original. The roadway is 10m wide and 15m above water level.

The bridge received its modern name during the pontificate of Gregory the Great (AD 590–604), who crossed the bridge to pray at the Vatican for an end to a plague raging through the city. According to the medieval legend which grew up around this event, he saw, in the sky above the mausoleum, an angel in the act of sheathing a flaming sword – which he interpreted to mean that God would listen to his prayers and bring the pestilence to an end. End it did and Gregory expressed his thanks to the almighty by renaming the mausoleum the Castle Sant' Angelo and the bridge the Ponte Sant' Angelo.

Outside Rome the best-known Roman bridge in Italy is the Ponte di Augusto at Rimini, greatly beloved of the sixteenth-century architect Palladio. He wrote of it: 'But considering, that of all the bridges I have seen, that at Rimini, a city in Flaminia, seems to me to be the most beautiful, and most worthy of consideration, as well as for its strength, as for its compartment and disposition. It was built, I judge, by Augustus Caesar.' Palladio made drawings of it from which it became a model for 'Palladian Bridge' construction in Europe in the later Renaissance period.

The particularly fine stonework of the Ponte di Augusto consists of a high-quality travertine, which has often been mistaken for marble. Nearly all the Roman bridges in Rome itself and the remainder of Italy were made of travertine, limestone or volcanic *tufa*. In the provinces, however, the use of other stones such as basalt and granites was not uncommon. Built around 20 BC and named after the Emperor Augustus, the Ponte di Augusto has five semi-circular arches of moderate span – up to 9m – and is unusual for a Roman bridge, especially a provincial bridge, in the amount of adornment attached to it. It features a strong cornice supported by a row of projecting dentils (tooth-like stone supports), while above each pier a panel framed by pilasters forms a classical pediment.

Although the spans are modest, the bridge commands special technical merit in so far as it crosses the river at an angle with the piers parallel to the riverbanks. It is thus a true skew bridge. Bridges of this type were unusual before the eighteenth century and required meticulous attention in the cutting of the *voussoirs* as well as a good understanding of solid geometry.

Built around 20 BC, the Ponte di Augusto at Rimini was much admired by Palladio; his drawings of it, emphasising its vertical lines, became a model for 'Palladian Bridge' construction in Europe.

Constrained by their rigid adherence to the semi-circular arch, the Romans nevertheless achieved great variety and beauty in many of their bridges: none more beautiful than the magnificent Alcantara Bridge across the Tagus River in Spain. Unusually, the name of the engineer, Caius Julius Lacer, is known, and his body lies interred close to the bridge. His epitaph reads: 'I leave a bridge forever to the generations of the world.'

The Romans built many important bridges in the provinces, notably in Spain – and none more majestic than the Alcantara (Arabic for 'The Bridge') Bridge over the Tagus River, with six spans ranging from 14m to 29m: the deck soars 50m above the river-bed on perfectly proportioned piers, 9m thick, pointed upstream and squared downstream. Built in AD 104, in the time of Trajan, its engineer, unusually, is known and accredited – Caius Julius Lacer – who was buried beneath a temple near one end of the bridge. He even allowed himself a triumphal arch, with an inscription in Latin saying: 'I have left a bridge that shall remain forever'. Not only does it still remain; it is one of the most magnificent bridges ever built.

The granite structure of the Alcantara Bridge has survived not only the swirling waters of many extreme floods, with some reaching above the springing level of the arches, but it has also suffered from military actions. In 1214 the Moors destroyed one of the smaller arches. One of the main spans, which had suffered damage in 1707 during the War of Succession, was deliberately destroyed by the French in 1812 to hinder the movement of the Duke of Wellington's army. Not to be deterred by such a trifling matter, Wellington's engineers threw a suspension bridge of ropes across the gap, allowing the Duke's siege artillery to cross.

The eastern fringe of the Roman Empire, long accustomed to the rise and fall of great civilisations, saw the most inglorious episode in the history of Rome in AD 260, when the powerful and rampaging Sassanian King Shapur I captured Valerian, co-emperor at that time with his son Gallienus. Installed as emperor by the troops of Gaul and Germany, Valerian persuaded the Senate that Gallienus should rule in the West, supported by the Danubian armies, while he, himself, should rule in the East, establishing his court at Antioch in Syria. He made a bad choice. Prolonged epidemic and Goths sweeping down from the Black Sea weakened his army and Valerian holed up in the fortress of Edessa near the upper Euphrates. Shapur talked him out of leaving its security to attend a peace conference and promptly took him prisoner. This news shocked the Empire. It was Shapur's finest hour.

Disillusioned by a lack of further successes, Shapur retired to his capital at Ctesiphon to occupy himself with large building projects. This enabled him to put his many Roman

In 260 AD the Sassanian King Shapur captured the Roman co-Emperor Valerian and enslaved him, possibly putting him to work with other Roman prisoners-of-war on this bridge at Shushtar, some 50km east of Susa. While there may have been Roman input into its construction, its function as both a dam and a bridge, together with its pointed arches, proclaim the strong eastern influence in its design. (From Hammerton)

prisoners of war to good use, notably in the construction of a combined dam and bridge spanning the Karun River at Shushtar, a town some 50km east of Susa in Iran. Valerian himself, treated as a slave by Shapur, may well have been forced to work on the 516m-long structure, much of which survives until today as a lasting reminder of Rome's shame. To make doubly sure his feat in capturing a Roman emperor would never be forgotten, Shapur had a carving made into a rock face at Naqsh-I-Rustam, near Persepolis, showing himself mounted on his war-horse with Valerian kneeling in supplication before him.

While Roman effort and perhaps technology, however unwillingly, went into the Shushtar Bridge, it does not have the characteristics of a typical Roman bridge. The pointed arches are quintessentially eastern, as is the combined purpose of the structure to act as both a bridge and a dam, which accounts both for its twisting alignment to cross the river following the line of the rock outcrop and for pier widths exceeding the arch openings. Roman touches include construction of the forty-one piers; each had a rubble core faced with masonry and arched openings in the spandrels to increase flood flow capacity.

The Iranians built many other bridges during the Sassanian period, some remains of which can still be seen. A notable example at Dizful in Khuzestan, built in the fourth century, has a total length of 380m. It drew particular notice for the first use by the Iranians of Roman coffer-dam techniques for siting the pier foundations. Elliptical arches, unusual in Persian construction, spanned the narrow 7m openings. Greatly restored, with many of the arches rebuilt in brick, it still stands. The Sassanids did not confine themselves to stone masonry bridge construction: known examples in kiln-burnt bricks include a bridge over the Dujayl River at Ahwaz and another over the River Tab in central Iran. Muslim bridge builders took over where the Sassanians left off, in some cases restoring or rebuilding ruined Sassanid bridges.

In a structural sense the masonry arch bridges built by the Sassanians did not match those of the Romans. The pointed arch, so common in Persian construction, although more flexible than the semi-circular arch in that the span:rise ratio could be varied, was otherwise, in theory at least, an inferior geometric form for bridge construction. In order to find a bridge of the immediate post-Roman period that not only matched the Roman structures but also outdid them in purity of design, it is necessary to look further east to China.

In its basic design concept the Great Stone Bridge, built between AD 605 and 617, and spanning the Chiao Shui River, placed the Chinese well ahead of the Romans and in advance of any European bridges built before the fourteenth century. Sited at the foot of the Shansi Mountains near the edge of the Great China Plain, near the town of Chao-hsien, it is regarded with great pride by the Chinese, and rightly so. Its segmental arch had two advantages over the semi-circular form from which the Romans rarely, if ever, departed. First, using a segment of a circle rather than the semi-circle allowed flexibility in the choice of a rise:span ratio, depending upon the angle subtended by the segment at the centre of the circle. Second, the self-weight and traffic loading on a bridge are carried in a more efficient manner by a segmental arch ring than by a semi-circular arch. High lateral forces are exerted by a segmental arch on its foundations and the stability of the Chiao Shui Bridge, still standing after 1,400 years, is a tribute not only to the merits of its structure, but also to the excellence of its foundations. It well deserves its name as the Anchi (safe-crossing) Bridge.

Structurally, the segmental arch of the Great Stone (Anchi) Bridge built in the time of the Sui Dynasty (AD 605–617) over the Chiao River, some 200km south of Beijing, was many centuries in advance of any bridge built in Europe.

The double arch spandrel openings at each end of the bridge lighten the loads on the foundations and increase the capacity for the passage of floodwaters. Equally important is their visual impact – by exposing the extrados they make very clear the function of the arch ring in conveying traffic loads and the bridge self-weight through the structure and into the foundations. In creating this remarkable 37.4m span with a rise of only 7.2m, the Sui civil engineer Li Ch'un enjoyed high regard in his own time, which has since become enhanced as a result of the lasting influence his structure has had on Chinese bridge design through the centuries. He also deserves the extravagant language used in the inscription carved on the bridge in AD 675, as described by Needham:

This stone bridge over the Chiao River is the result of the work of the Sui engineer Li Ch'un. Its construction is indeed unusual, and no one knows on what principle he made it. But let us observe his marvellous use of stonework. Its convexity is so smooth, and the *voussoir*-stones fit together so perfectly . . . How lofty is the flying arch! How large is the opening, yet without piers! Precise indeed are the cross-bondings and joints between the stones, masonry blocks delicately interlocking like mill wheels, or like the walls of wells; a hundred forms organised into one. And besides the mortar in the crevices there are slender-waisted iron clamps to bind the stones together. The four small arches inserted, on either side, break the anger of the roaring floods, and protect the bridge mightily. Such a masterwork could never have been achieved if this man had not applied his genius to the building of a work that would last for centuries to come.

Chinese bridge builders achieved forms of stone arch construction generally lighter than that of their Roman counterparts, the typical Chinese example according to Fugel-Meyer consisting of a thin stone shell loaded with loose filling confined by side walls and topped with stone slabs to form the deck. The Anchi Bridge consists of a stone *voussoir* shell made up of twenty-eight rings placed side by side, each acting independently so that damage to one ring has no effect on the others or on the overall stability of the bridge.

4

TRAVERSING THE LAND

Most early ancient civilisations were concentrated in a fringe area around the Mediterranean, but their influence spread far beyond this area, mainly through trade. Vestiges of some of the old trade routes survive to this day. These routes were not roads in the modern sense; they often consisted of little more than trails or rough tracks, sometimes obvious and well worn, sometimes barely discernible unless the traveller knew the landmarks to guide him. As such routes were generally unsuitable for wheeled traffic, loads were mostly carried by humans or animals.

From northern Europe came amber, used for statuary and jewellery as well as being an important ingredient in primitive medicine. There are known to have been at least three major amber routes through Europe. One ran southwards from present-day Hamburg to the Rhine near Frankfurt and then followed the Rhine to Basle, where it turned westward to meet the Rhone, which it followed to the Mediterranean. Another followed the Elbe from Hamburg to Magdeburg, then southwards to the Danube and over the Brenner Pass to Venice. Yet another started in East Prussia where amber is still to be found and crossed the rivers Vistula, Oder and Danube to reach the northern tip of the Adriatic.

The amber trade appears to have had its origins at least as early as 2000 BC; although it came to an end in Roman times, remains of the old amber routes are still found. Traders carrying valuable goods were a natural prey for local tribesmen, and their graves lining the routes have yielded valuable information to archaeologists. Traces of corduroy roads have been found, where the routes crossed wet or swampy land to avoid long detours. These consisted of up to four layers of oak or alder logs laid criss-cross, alternately lengthwise and crosswise.

Among the most popular commodities sought by the Mediterranean peoples were the aromatic gums and spices, on which southern Arabia had a monopoly. Trains of camels carried aloes, balsam, myrrh and incense across the Arabian deserts to Mediterranean ports for distribution throughout the known world. The incense routes followed the wadis and passes and gave birth to many important towns which provided rest, sustenance and, no doubt, entertainment in a variety of forms for the weary or simply bored travellers. Stiff fees – in effect tolls – charged for traversing lengths hacked with great effort through mountainous terrain enabled local governments in southern Arabia to make rich pickings from the passing caravans. Transit towns such as Marib and Petra became rich through the incense trade.

The Silk Road to China was the longest of the ancient trade routes. Great bales of Chinese silk passed through the Yumen Gate in the Great Wall of China and headed westward carried by caravans sometimes consisting of hundreds of camels. At Dunhuang the route split in two to cross the great Tarim Basin, the route chosen usually depending upon the vagaries of the Tarim River, which frequently changed its course. Both roads skirted the great deserts and salt marshes of central Asia and climbed high mountain passes to join up again at Kashgar. On the southern route a series of walled oasis cities provided the travellers with the possibility of overnight rest and protection and the replenishment of water supplies. They could look forward to a longer stopover and reprovisioning on reaching Shan-shan (Cherchen), the first major city on the southern route and 1,000km south-west of Dunhuang. Then on to Khotan, where goods could be traded for jade found in local river beds, and Charkand, traversing the southern edge of the forbidding Taklamakan Desert for a distance of another 1,000km; this journey took several weeks or months. In some cases the caravans overwintered at one of these major towns rather than brave the high Pamir tablelands to the west. After weeks or months negotiating the barren Tarim Basin, the Charkand River and the wheat fields of Charkand must have seemed like an enchanted land to man and beast.

Travellers taking the northern route had to cross the soulless salt beds of the Lop Desert and then negotiate salt marshes of Lop Nor before reaching the first major city, Loulan, beyond which they faced more sandy desert before arriving at Korla, situated in a well-watered valley with good grazing. By the third century AD drier conditions led to the abandonment of this part of the northern route and forced the travellers to head north from Ansi, before reaching Dunhuang, taking a wide loop to the north of the Great Wall and across the Gobi Desert to rejoin the old route at Korla. Their onward route then took them to Kucha, approximately midway along the older route and an important local administrative centre, and Aksu, another major town on a tributary of the Tarim River, and finally to Kashgar, the westernmost point of the Tarim Basin and the meeting point of the southern and northern routes. Most traders chose to travel in large caravans of

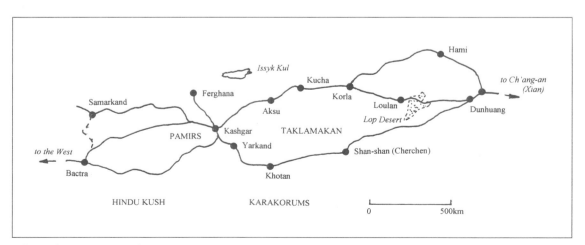

Silk Road routes in Central Asia.

hundreds of camels, organised by caravan masters, and accompanied by armed guards as protection against groups of local bandits, some of whom came in official disguise. Towns like Hami, Korla and Aksu exacting hefty taxes in cash or kind in return for supplying shelter and allowing passage. They also made exorbitant charges for supplies. One blessing arising from the harsh natural conditions, as noted by Marco Polo, was the absence of any dangerous wild animals, owing to the lack of anything for them to eat!

West of Kashgar a number of possible routes climbed initially through the steep, and often precipitous, eastern ramparts of the high Pamir Mountains. On reaching the watershed they were past the worst and faced easy descents across gently sloping plateaus and river plains to the great trading centres such as Ferghana and Samarkand, the fabled city of the Arabian nights, where the silk bales were exchanged with western traders for woollen textiles, pearls, glassware and, from Arabia, spices. A more southerly route starting from Kashgar or Charkand headed south-west and west to Bactra at the head of the Indian Grand Road, giving access to the Punjab and the rest of India, a route whereby Buddhism entered China. Alternatively, the travellers could head west to the great caravan town of Merv, whence the road to the west ultimately linked up with the Royal Road in Persia and gave access to towns such as the port of Antioch and, from there, shipment to other ports of the Mediterranean civilisations.

In the sixth century BC, for military, trade and communication purposes, the Persians established a road system, most of which resembled the old trade routes in that little attempt was made to construct paved roads. The work done in making these roads was the minimum required to consolidate the empire. Rivers were usually crossed at fording points or by ferry or, in some cases, by floating bridges; some cutting and levelling would have been required over mountain passes and building up of the road surface over swampy areas. Clearing would have been required in forested areas. These roads extended throughout Asia Minor, Babylonia, Egypt, the Caucasus and Caspian regions, Persia itself and probably parts of India. The most famous of all, the Royal Road, created at the order of the Persian King Darius who came to power in 522 BC, ran for 2,700km from Sardis, capital of Lydia and now part of modern Turkey, to Susa near the head of the Persian Gulf. Sufficiently aware of its existence to be able to give a description of it, Herodotus may well have made good use of it in his travels:

> At intervals all along the road are recognised stations, with excellent inns, and the road itself is safe to travel by, as it never leaves inhabited country . . . the total number of stations, or posthouses, on the road from Sardis to Susa is 111. If the measurement of the Royal Road in parasangs is correct, and if a parasang is equal (as it indeed is) to 30 furlongs, then the distance from Sardis to the palace of Memnon (450 parasangs) will be 13,500 furlongs. Travelling, then, at a rate of 150 furlongs a day, a man will take just ninety days to make the journey . . . I would point out that the distance from Ephesus to Sardis should be added to the total, so that one gets, as a final measurement from the Aegean to Susa – the 'city of Memnon' – 14,040 furlongs.

A travelling rate of 30km per day assumed by Herodotus, by foot or possibly by donkey, as would be the case for the ordinary mortal, may seem rather modest, but owing

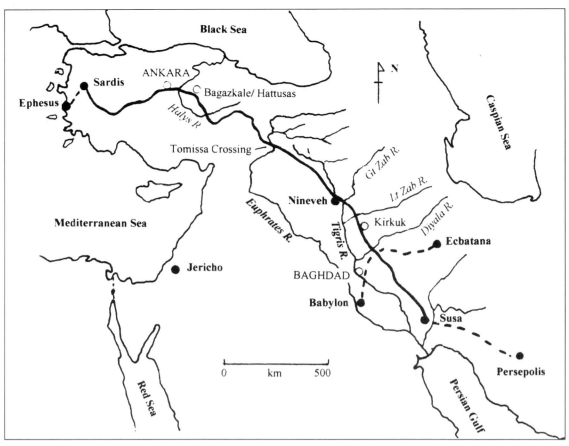

The Persian Royal Road ran from Sardis to Susa, a distance of 2,700km. Herodutus estimated that a man could traverse it in ninety days, but that messengers on horses could do it in ten days.

to poor ground conditions and frequent flooding the road had to be diverted from the Tigris Valley to cross the precipitous Zagros Mountains in western Iran, climbing to an elevation of 1,500m. Other obstacles which had to be surmounted included the Taurus Mountains, the 1,980m-high Karadja-Dagh and the narrow Tigris Valley at Mosul.

Darius ordered the road to be built mainly to assist in keeping his vast empire under control. Rapid communications were essential for this and the road enabled messages to be passed from Susa to Sardis in ten days, by having staging posts at convenient intervals with fresh horses and riders. No doubt the ordinary citizen on foot or donkey learned to vacate the road quickly on hearing the pounding hooves of these flying messengers. According to Herodotus, nothing in the world travelled as fast as these Persian couriers.

Ironically it was in part the excellent Persian road system, and the well-structured system of government, which helped make it possible for Alexander the Great not only to defeat the Persians, but also to establish, within a few years, an empire extending from Macedonia in the west to the Indus River in the east and Egypt in the south. The roads facilitated rapid movement of his cavalry, heavy infantry and war machines. Having assumed the throne of Macedon in 336 BC, he quickly crushed a Greek rebellion before invading Persia and inflicting three major defeats on the Persians between 334 BC and 332 BC, after which

Persian resistance crumbled. These victories enabled Alexander to assume rapid control of the widespread empire by replacing local satraps, or governors, with his own appointees. He then turned his attention to consolidating and extending his empire.

Having finally defeated the Persians in 330 BC, and annexed a ready-made empire, Alexander continued to march eastward, crossing high snow-covered mountains and searing deserts against fierce resistance to reach the Indus River four years later in 326 BC. Leaving garrisons behind to consolidate these gains, Alexander followed the Indus to the Indian Ocean, then doubled back to Babylon, where he died in 323 BC. His generals immediately divided his empire between themselves, Seleuces establishing the Seleucid Empire in Asia. The garrisons left behind on the Indus by Alexander quickly compounded before the northern Indian king, Chandragupta Maurya, who massed a huge army in 305 BC to repel an invasion by Seleuces. The Greek general thought better of his aggressive intentions at the last moment and reached an amicable agreement with Chandragupta. With northern India now under his control, Chandragupta went on to capture a succession of southern states, a process completed by his grandson Asoka, under whose reign (274–236 BC) the Mauryan Empire encompassed virtually the whole of modern India and Pakistan, with the exception of the southern tip of India.

Strong and stable empires inevitably produce outstanding civil engineering works, and the Mauryan Empire proved to be no exception. Administrative departments were set up for the construction of harbours, public buildings, sanitation and water supply systems, canals and roads. Chandragupta had his own Royal Road built to link his capital Pataliputra on the Ganges with the north-western outpost of the empire, this road later becoming the main east–west arterial way through northern India, with a total length of over 4,000km. As with many successful and far-sighted monarchs throughout history, Chandragupta had the wisdom to appoint an outstanding minister, Kautilya, the author of *Arthasastra*, a treatise advising a would-be king on how to gain and hold power. *Arthasastra* was something of a forerunner to Machiavelli's *The Prince*, written 1,800 years later. Kautilya's advice on statecraft apparently included the establishment of a widespread spy network reporting to the king. These spies picked up much of their information from the motley cross-section of people travelling the roads, but the roads served the even more important function of ensuring that the information could reach the king quickly by fast courier.

Much of what is known today about the Mauryan period comes from fragmentary accounts left by the Greek, Megasthenes, the Seleucid ambassador to the Mauryan court. He was particularly impressed by the excellence of the road system, with its provision of signposts at road junctions, distance markers every ten *stadia* (about 1 mile or 1.6km) and rest-houses for travellers and pilgrims at regular intervals, equipped with wells, which must have been a welcome sight after a long and hot dusty journey.

Paved sections of the road were confined to the cities or approaches to the cities: the paving materials used, according to availability, included stone slabs, fired brick, lime-stabilised soil and gravel. Most roads consisted of a formed profile with side ditches for drainage and a compacted soil surface. Megasthenes gives a brief description of their construction. After the soil along the road alignment had been studied and tested by experts, workmen moved in, armed with picks, axes and scythes, to cut down creepers, undergrowth, bushes, thickets and trees, move or flatten rocks, prise up tree stumps, level slopes and humps, and finally fill up holes and depressions to give a solid, even road surface. The road surface was raised above the level of adjacent land, which, together with the side ditches, ensured good drainage. Despite these provisions heavy monsoon rains falling on the unpaved roads would have made constant and expensive maintenance necessary. The State raised the necessary funds by exacting tolls from merchants using the imperial or royal highways. This payment also supposedly brought them some protection by the State against robbers who preyed on travellers. Exacted at every town along the route, the tolls amounted to a levy of between one-twentieth and one-tenth on cloth, oil, pottery, sugar, grain and other merchandise. Merchants trying to avoid the levees by resorting to side-roads skirting the towns ran the gamut of the efficient spy system and risked being thrown into jail and having their goods confiscated.

The roads Alexander found in the areas that had been under Persian domination represented a vast improvement on the roads in Greece itself, many of which were little more than tracks, barely suitable for pack animals, and sometimes following stream beds that were dry most of the year. A number of factors contributed to the poor road system in Greece, including the lack of political unity in the country, the mountainous nature of much of the topography, and the prowess of the Greeks as a sea-faring people. The strongly incised coastline of Greece gave ready access from many parts to the sea. The roads and tracks that did exist mostly connected the villages or towns, or perhaps quarry sites, to the navigable point on a river or the coast. Even within the cities the roads often consisted of little better than undrained, unpaved, refuse-strewn alleys, muddy in winter, dusty in summer, and perpetually smelly. Not surprisingly the mud-brick houses were built with blank walls facing the streets and with the rooms opening out into an inner courtyard.

Although it was perfectly acceptable for ordinary citizens plying their trade to travel in discomfort and with considerable difficulty, the gods themselves, together with their appurtenances and worshippers, could not be expected to put up with such conditions. Thus the exceptions to the generally poor road conditions in Greece were the 'sacred roads' built to connect the centres of population to the important religious sanctuaries, notably Delphi. These roads had to take the lumbering four-wheeled carts used to convey the religious offerings, accoutrements and statues of the gods without severe jolting. The roads tended to follow the contours to keep gradients to a minimum, some consisting simply of parallel ruts spaced 1.38–1.44m apart, commonly 70–100mm deep and 200–220mm wide. On a particularly rough stretch of road the depths of the ruts could be increased to as much as 300mm in order to maintain a level track for the wheels to run on. The ruts were formed by cutting into rock or hard stony soil, or by embedding pre-cut rutted stone blocks into soft ground. The spacings of these ruts correspond remarkably closely to the standard rail gauge today of 1.43m.

These rutted roads were not unique to Greece; even earlier examples have been found in Malta and Italy, in the latter case of Etruscan age. As most of the roads had only a single pair of ruts, when two carriages approached each other one had to give way. Manoeuvring one of these ponderous carts out of the ruts would have been no easy task. It would have led, in some cases, to violent quarrels, such as that which resulted in Oedipus killing his father, a stranger to him, on the Cleft Way to Delphi. In rare instances, passing places with duplicated sets of ruts were provided.

The Assyrian kings, Sargon II (721–705 BC) and his son Sennacherib (704–681 BC), had magnificent processional ways built in the alternative capitals of Assur and Nineveh. The processional road to the Ishtar temple of Assur had a base of burnt bricks separated by bitumen layers, resting on a stone and gravel foundation, and surmounted by paving slabs of red-veined gypsum with bitumen-filled joints. Kerbstones defined the edges of the road and grooves gouged into the flat paving slabs guided the wheels of the ceremonial wagons. The grooves may have been partly filled with a softer substance such as straw to give a smoother and quieter ride. A processional way at Nineveh, built by the order of Sennacherib, was 27m wide and lined with lofty pillars. Severe punishment by impalement awaited anyone who lessened this road in any way, perhaps by building a balcony

Minoan paved road near Knossus. A system of paved roads existed in Crete shortly after 2000 BC, connecting Knossus to other important towns.

overhanging it or even by parking a cart or chariot along its length. Perhaps some modern planning officials would like to have the powers of punishment afforded to their ancient counterparts.

The processional way in Babylon built by Nebuchadnezzar to run from the great Ishtar Gate through the centre of the city owed much of its design to the earlier Assyrian examples. It consisted of a base of flat burnt-clay bricks or tiles, set in bituminous asphalt, with a surface paving of flat limestone flags. Two rows of stone lions fringed the 18m-wide road. When some of the limestone blocks were removed in modern times they revealed an inscription by the king stating: 'I had the Road of Babel laid with Shadu stone flags for the procession of the great Lord Marduk', thus defining very clearly the purpose of the road.

In most of the ancient world paved roads or sections of roads were confined to processional ways or special roads within or leading into cities. One exception to this was in Minoan Crete, where a system of roads in existence shortly after 2000 BC linked the capital Knossos to other major cities around the coast. Some lengths of these roads compared in workmanship with those of the Romans two millennia later. One major road from Phaestos in the south to Knossos in the north had to pass first through the Messara Valley, before climbing over 1,500m to cross the precipitous mountain spine dividing the north and south of the island. In a typical paved cross-section the Minoans carefully excavated and compacted the road base and, on this, placed a 200mm-thick layer of rubble and broken stones in clay to make it watertight, then 50mm of clay, into which was embedded, typically, a 4m wide surfacing of basalt flags flanked by mortared pieces of limestone. The Minoans knew of the need to keep water away from the subgrade and, in addition to the watertight layer immediately above the subgrade, they provided stone side drains to carry away surplus surface water.

Domination of the Mediterranean required a mastery of the sea and the Romans resolutely achieved this, although they seem to have been much more at home on land than on the sea. They perfected road construction primarily to serve their military needs, but the network eventually found increasing use for general travel, communications, trade, and even tourism. By the time of the Empire some 80,000km of major roads had been constructed and perhaps twice this length of lesser roads. It was possible to travel on a major road from Bordeaux to Jerusalem.

In the eighth century BC Greek colonists established the settlement of Cumae in the Bay of Naples, attracted by the rich volcanic soil of the plains of the Campagna lying between modern Naples and Rome. The settlement became a thriving centre for the sale of grain grown on the plains, and enabled the Greeks to extend their hold over all the southern portion of Italy and over Sicily. They employed sophisticated building techniques in masonry and used lime mortar in the buildings and in some of their street construction. Herodotus himself spent his last years here and died in Thurii around 420 BC. It may have been from these Greek colonists that the Romans learned to use lime mortar; but it appears to have been the Romans themselves who developed a hydraulic cement by combining lime with a volcanic earth (now called *pozzuolana* after Pozzuoli, near Naples, where this substance occurred). This cement had cementitious and strength properties comparable with modern cement. The Romans found similar volcanic deposits in various

The Via Appia, built by the blind engineer Appius Claudius from 312 BC to 295 BC, originally ran 210km from Rome to Capua, but was later extended to the port of Brindisi. Ornate tombs, milestones and lavish villas flanked the road. In 71 BC the 6,000 followers of Spartacus were crucified on crosses erected alongside the road.

parts of their extensive empire and commonly incorporated them into their road construction.

The area stretching over 300km to the north of Rome, with the Arno River as its northern boundary and the Apennines its eastern boundary, was the homeland of the highly talented Etruscans, a people who may have had their origins in Asia Minor. Essentially an urbanised people, they lived in perhaps a dozen flourishing city-states interspersed with smaller towns or villages. Well-drained, gravelled roads connected these urban centres, establishing a network later absorbed into the Roman road system in Italy. Within some of the towns the main streets were paved, well drained and up to 15m wide, with sidewalks. The Romans are known to have rebuilt some of these and, in doing so, acquired at least some of the knowledge of road construction that they put to such good use over the length and breadth of their empire. The provision of stepping stones to protect pedestrians against water coursing along the paved streets between the raised pavements, which can be seen in Pompeii and Herculaneum today, derived directly from Etruscan practice.

It can be safely assumed that Etruscan influence led to the early steps of the urbanisation of Rome. Its progress towards becoming the most important city in the world received a setback in 390 BC when it was sacked by the Gauls, but it recovered quickly enough. Construction of the Via Appia, the first paved road of any substantial length in the world outside Crete, began in 312 BC. It was, in fact, initially gravelled, but the paving had been added by 295 BC. The first section, completed by the engineer, the blind Appius Claudius, after whom it was named, ran from Rome to Capua, a distance of 210km. It served to bring the rich harvests of the Campagna into Rome. The following century saw it extended in a south-easterly direction to Taranto at the top end of the heel of Italy and from there to Brindisi on the east coast. A short trip by ship across the southern end of the Adriatic Sea linked this port with Dyrrhachium (now called Durres) on the coast of modern Albania. From Dyrrhachium, another of the great Roman roads, the Via Egnatia, climbed into the high mountains which form the present border between Albania and Montenegro, then traversed difficult country of alternating high mountains and deep valleys across central Macedonia, and passed through the ancient towns of Edessa, Thessalonika and Philippi to its terminus at Byzantium on the Bosphorus. Originally built for military purposes, it later became used almost exclusively by civilian traffic.

The Romans were not averse to incorporating more ancient road systems into their construction programmes and the route they chose for the Via Egnatia followed, to a significant extent, the Kato-oclos (lower road) route that existed in the time of Herodotus and was used by Darius, Xerxes and Alexander. These more ancient routes tended to avoid natural obstacles such as unstable or marshy ground, heavily forested areas or deep

The Via Egnatia, originally built in Republican times for military purposes, climbed into the mountains bordering northern Albania, then traversed central Macedonia to its terminus at Byzantium. It eventually became used almost exclusively for civilian traffic, as with many Roman military roads.

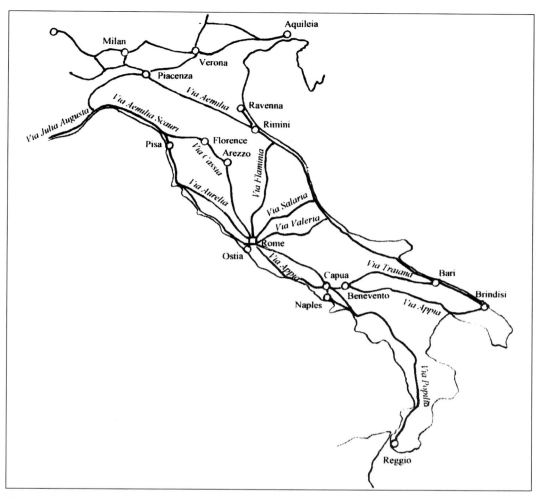

The roads that led to Rome: principal Roman roads in Italy.

ravines, whereas the Romans, applying more purposeful civil engineering methods and more advanced technology, favoured the direct route where practicable, with less concern about taking on such difficulties.

If they considered a route important enough they would employ their civil engineering skills to shorten it by cutting or tunnelling through high ground, constructing causeways across marshes and throwing bridges and viaducts across valleys and ravines. The Forlo Tunnel on the Via Flaminia near Pesaro, cut in AD 77 by order of the emperor Vespasian, with an opening 6m high and 5.5m wide cut through 40m of solid limestone, is still used by modern traffic. A much longer tunnel still in use passes under the Pausilippo between Naples and Pozzuoli; built in 36 BC, it has a length of 1,460m. Seneca makes particular mention of the rising dust in the tunnel, which blotted out the light and must have made travelling through it an unpleasant experience. A notable example of a deep cutting occurs on the Appian Way near Terracino where a 36m-deep excavation enabled the road builders to negotiate a high marble cliff.

Once the Via Appia had been extended to Brindisi, travellers from Rome to this port could embark not only for Via Egnatia, but also for other ports all round the Mediterranean. The importance to the Romans of one of these ports, Antioch, can hardly be overstated. While assuming great importance in military operations against the Parthians, descendants of the Persians and doughty foes who occupied most of modern Iran and Iraq, it was also the focal point through which enormous amounts of merchandise, including incense and spices and other goods from Arabia, passed through on their way to Rome. Roads radiated from Antioch to the Euphrates and Tigris, to the Red Sea and on to Pelusium in Egypt, to the Persian Gulf and to all the important cities of Asia Minor.

For centuries travellers out of Rome heading for the south along the Via Appia have been gently lulled into a time warp by the ornate tombs, milestones, monuments and opulent villas flanking the road, and by glimpses of ancient aqueducts in the distance. It witnessed, however, one of the most grisly events in the history of mankind. Professional gladiators, joined by slaves, under the leadership of the Thracian, Spartacus, terrorised and plundered the rich Italian countryside for three years from 74 BC to 71 BC. The authorities in Rome entrusted the task of subduing Spartacus and his followers to the ex-praetor, Crassus, a wealthy wheeler-dealer who included slave trading in his multifarious moneymaking activities. With a highly trained force of 40,000 men Crassus finally cornered and killed Spartacus in Apulia, in the south of Italy. He then erected 6,000 wooden crosses along the Appian Way and, on each one, crucified one of the erstwhile followers of Spartacus. Crassus went on to become elected by the Assembly in 70 BC to joint Consul of Rome with the more military-minded Pompey.

Another famous road out of Rome built during the Republican period was the Via Flaminia, which crossed the Tiber by way of the famous Milvian Bridge, then followed the valley of the Tiber, taking a route almost due north over the Apennines to the Adriatic coast. Here it changed direction to take a north-westerly route along the coast, passing through Ariminum (modern Rimini). Shortly after Ariminum it forked, one leg continuing along the coast to Ravenna and the other, the Via Aemilia, striking inland to Placentia (Piacenza) on the River Po. Placentia lay at the heart of Roman northern Italy and from here major roads met and radiated to all quarters of the Roman world. Ironically, its eponymous builder Gaius Flaminius, consul and army general, fell in the fighting in 217 BC against Hannibal, a few years after completion of his road. Hannibal used the Flaminian Way for his advance against Rome; but he wisely changed his mind during the march and by-passed the city in favour of the rich grain-lands of the Campagna and the ports of southern Italy. Unable to get any reinforcements from Spain, Hannibal nevertheless held out in southern Italy for four years against a series of sieges by the Romans until, with his influence rapidly declining, he returned to Carthage to fight further futile battles against the Romans.

The most important province of Rome was Spain, which provided, among other things, olive oil, a staple agricultural product used by the Romans for eating, cooking, lighting and as a substitute for soap. The fertile valleys of the Guadalquivir River in southern Spain produced the highest-quality olive oil in the ancient world. Other important products flowing into Rome from Spain included grain, vines and livestock, the last including Lusitanian horses, unrivalled for their speed. Mineral wealth also abounded in Spain to be exploited by the Romans: gold in the north-west of the country, silver near Cartegena which the Carthaginians had mined previously, copper in the south-west, particularly along the Rio Tinto River, where it is still mined. Lead, tin, iron and mercury were other metals mined by the Romans in Spain.

In order to acquire Spain and its abundant riches of agricultural products and raw materials, the Romans had first to overcome the Carthaginians, who had a stronghold with their base at Cartegena, then known to the Romans as Carthago Nova. Two important victories by the armies of Scipio – in 209 BC and 206 BC – broke the Carthaginians, but it took another seventy years to subdue the defiant Celtiberean and Lusitanian tribesmen and finally secure Spain for Rome. In the following century most of the fighting in Spain took place between rival Roman strongmen battling for the rich spoils of the country; the great rivals Caesar and Pompey fought two battles on Spanish soil, Caesar winning both to ensure his position as supreme leader of the Roman world.

Augustus finally brought peace to Spain and he set about an active programme of road construction to pacify and control the province and to exploit its riches to the full. Access by road to Spain, at this time, was principally through Vada Sabatia on the Ligurian coast, which could be reached either from Placentia by the Via Claudia Augusta or directly from Rome along the Via Aurelia, which passed through the ancient Etruscan towns of Caere and Tarquinii before heading for the Ligurian coast. From Vada Sabatia the road followed the Mediterranean coast to skirt past the ancient Greek port of Massilia (Marseilles) to Arelate (Arles) and nearby Nemausus (Nîmes), whence it followed the coast to Valentia (Valencia) and Carthago Nova. Augustus had a road system constructed which encompassed the perimeter of Spain, with offshoots penetrating inland to the mining areas, along which the ores and metals could be taken to the seaports for shipping to Rome. He also initiated an extensive road system in Gaul based on Lugdunum (Lyons) as the nodal point, and this gave a further access to Spain from Tolosa (Toulouse), which crossed the mountains to Caesaraugusta (Zaragoza). While primarily serving Gaul itself, which had a thriving economy in agriculture and in industrial production, particularly pottery and glass, the road network in the province also played a major part in enabling Claudius, in AD 43, to muster an army of 40,000 men at Boulogne for his assault on Britain. Claudius himself marched from Marseilles to the Channel coast.

As both a sea port and a river port, and with easy access to the European mainland, Londinium (London) swiftly became the main trading centre in the country, with many roads radiating from it: the south-west road to Durnovaria (Dorchester) also gave access to Calleva (Silchester), Venta Belgarum (Winchester), Sorbiodunum (Salisbury), Glevum (Gloucester), Aqua Sulis (Bath) and South Wales; the road to Deva (Chester) passed through Verulamium (St Albans), and beyond Chester access could be gained to North Wales, Lancaster and the Lakes; Ermine Street ran due north to Lindum (Lincoln) and then on to York, Newcastle and Hadrian's Wall. Lesser roads linked London to the early capital, Colchester, and to the Channel ports. At Lincoln the north road met the great Fosse Way, which cut across several of the roads radiating from London as it struck south-west, in a direct line from Lincoln to south Devon, passing through Ratae (Leicester), Corinium (Cirencester), Bath and Lindinis (Ilchester). Originally constructed as a military system to subdue and pacify the country, these major roads continued to serve the military garrisons by ensuring a rapid means of communication between them and by providing a ready access for supplies. Eventually these roads were supplemented by a secondary road system to satisfy a growing trade in agricultural products, raw materials and manufactured goods. It has been estimated that the Romans built nearly 10,000km of road in the first forty years of their occupation.

After sacking Carthage in 146 BC and destroying the city, the Romans progressively expanded their possessions in Africa. But three and a half centuries were to pass before the campaigns of Septimus Severus (himself born in Leptis Magna in Tripoli) led to an entire strip, averaging some 160km in width along the North African coast from Morocco to Egypt, becoming absorbed into the Roman Empire. Within the area previously under Carthage an excellent road system already existed to exploit the rich agricultural products of the country. The Romans expanded this system, originally for military purposes, but with the subsequent addition of later roads for civilian purposes the entire system totalled some 15,000km in length, linking the coastal road and ports to inland communities. While grain comprised the main commodity flowing along these roads, much of it destined for Rome, other agricultural products also found their way into the markets of that city, including olives and more exotic fruits such as figs, pomegranates, almonds and dates. Exotic animals, too, existed at that time north of the Sahara – lions, leopards, antelopes, ostriches and some elephants – which the Romans hunted for sport and for capture and sale. Captured lions were caged and taken by road to the nearest port for shipment to Rome to satisfy the bloodthirsty appetites of the spectators at the Colosseum.

As well as marching troops, other pedestrians and pack animals – donkeys, mules and horses – Roman roads were designed to take a variety of wheeled vehicles such as passenger-carrying carriages drawn by mules or lumbering solid-wheeled ox-drawn carts. The fastest-moving traffic belonged to the Imperial Post system using runners, who lived their lives on the road, or horse-mounted messengers distinctive in their flat hats, woollen cloaks and 'barbarian'-style breeches, an odd mode of dress for Romans, who usually preferred to display bare legs. Perhaps they saw themselves as the direct descendants of the Persian 'flying messengers' on the Royal Road. Reminiscent of the Royal Road itself, relay stations provided fresh horses at intervals of 15–20km, depending on terrain, enabling messengers to achieve 150km a day or even more in emergency. Hostelries offering accommodation, food and drink, as well as stabling for horses and workshop facilities to repair damaged vehicles, were provided at intervals of about 35km or one day's travel.

Wealthy Romans liked to journey with all the comforts of home, travelling in state in their own capacious vehicles, followed by other carts carrying their baggage and servants. Eschewing the lower life (and lowlifes) catered for by the hostelries, they pulled off the road at a convenient place and, tended by servants or slaves, enjoyed a good meal before retiring to the beds set up in the carriages or in tents. Highway robbers constituted a constant hazard, sometimes working in collusion with innkeepers, who tipped them off if travellers stopping overnight had valuable possessions with them. Along some stretches of roads, robber encampments grew up and became complete family settlements. Perhaps the most famous and feared chief was Bulla, who, with his 600 men, mostly runaway slaves, terrorised the Via Appia for two years during the third century AD. His spies kept him

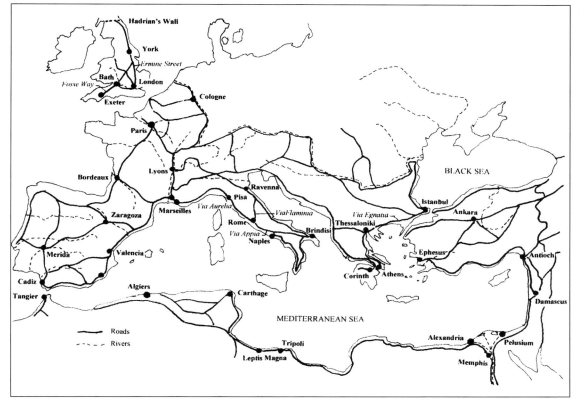

Principal roads of the Roman Empire.

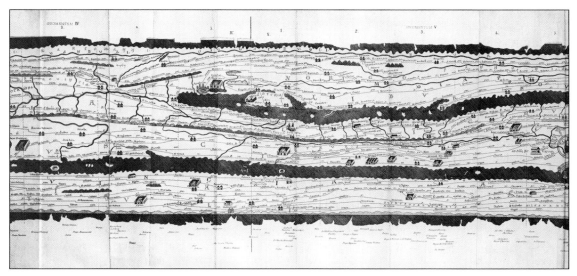

This Medieval copy of the compressed and distorted Peutinger Table, 6.8m long and 340mm wide, shows the entire system of Roman Imperial Post Roads, together with various features, buildings and road distances. This segment has Italy in the centre, with Rome at the right-hand edge. Africa and the Danube countries appear, respectively, below and above Italy.

informed of all important and wealthy travellers setting out from Brindisi or Rome; but eventually his mistress betrayed him to the Imperial Cavalry, which led to his capture and ultimately to his being fed to the wild beasts in Rome.

In order to make efficient use of the vast network of Roman roads stretching from Scotland almost to India, the traveller would have needed road maps or guides indicating routes, distances and geographical features such as rivers, main towns, wayside inns and hostelries. These existed, although no original copies appear to have survived. A survey of Spain, Gaul and the Danube provinces made in the time of Augustus was sculpted on marble and displayed near the Pantheon in Rome. Another marble map, ordered by Vespasian (AD 69–79) for public display, showed the fourteen quarters of Rome. Parts of this still survive. Once on the road the traveller enjoyed the benefit of good signposting and milestones, up to 2–4m in height, showing the distance from the nearest town in Roman paces, or actually double paces equal to 1.48m. Information other than distances commonly inscribed on a milestone might include the name of the official responsible for building or restoring the road, the date of the work, the name of the emperor, with all his titles, the source of finance for constructing the road, particularly if privately funded, and occasionally engineering or other works associated with the road. Even words of welcome might be addressed to the traveller, which could well have had the intention, and indeed the effect, of reinvigorating flagging footsteps.

Two records of the road system that have survived, although not in the original, are the Antonine Itinerary and the Peutinger Table. The former is a tabulation of major roads in the Empire, with place names and distances, apparently based on a wall map dating from the time of Caracalla (AD 212–217), although the Itinerary itself seems to have been drawn up rather later, perhaps between AD 280 and 290, in the time of Diocletian. The Peutinger Table is in map form, compressed and distorted to fit the entire system of Roman Imperial Post roads, and extending even beyond the frontiers of the Roman Empire into Iran and India, recorded onto a sheet 6.8m long and 340mm wide. It gives no indication of relative positions or orientations of the roads. Road forks and crossroads are shown, together with symbols representing various structures and amenities such as forts, temples, cities, towns, ports, lighthouses and spa establishments. Road distances between various places are shown. The existing map, in the Library of Vienna in Austria, is a mediaeval copy found towards the end of the fifteenth century and given to Conrad Peutinger of Augsburg, from whom it derived its name.

No detailed descriptions of Roman road building appear to have survived, but some idea of the activity involved can be garnered from the pen of the poet Statius (AD 48–96), who applied his talents to producing an eloquent account of construction of the Via Domitiana which ran from Mondragone in the Campagna to the nearby port of Pozzuoli. The technical detail in the description points to his being a keenly observant bystander:

The first labour was to prepare furrows and mark out the borders of the road, and to hollow out the ground with deep excavation; then to fill up the dug trench with other material, and to make ready a base for the road's arched ridge lest the soil give way and a treacherous bed provide a doubtful resting-place for the o'erburdened stones; then to bind it with blocks set close on either side and frequent wedges. Oh! How many gangs are at work together! Some cut down the forest and strip the mountain-sides, some plane down beams and boulders with iron; others bind the stones together, and interweave the work with baked sand and dirty tufa; others, by dint of toil, dry up the thirsty pools, and lead far away the lesser streams.

It is likely that the furrows were ditches used both to mark the boundaries of the road and to drain the subgrade. Other features are clearly described: the cambered road surface to shed water falling on the pavement, to prevent it penetrating into the road base and subgrade; a deep excavation to expose a sound subgrade and provide adequate depth of built-up pavement; kerbstone blocks and wedges to hold them firm; the breaking up of boulders to make crushed stone suitable for inclusion in the pavement and binding the stones with lime (baked sand) and *pozzuolana* (dirty *tufa*). Emphasis is given to the need to keep the work dry. The whole passage conveys a scene of great activity, almost giving the reader a feeling of witnessing the work at first hand.

A typical important Roman road had a thickness of about 1.2m and a width between kerbstones of 5–9m. A trench about 1m deep was dug to the full width of the road and a layer of sand, 250mm thick, spread on the soil subgrade and covered with a skim of mortar. Alternatively, a course of flat stones might be used as the base course. Immediately above the base course a layer about 400mm thick of large broken stones or cobbles mixed with water was usually favoured, heavily tamped to provide a solid bearing. The third layer, which brought the construction up to about natural ground level, consisted of a 300mm thickness of crushed stone concrete, possibly cambered, and supporting a pavement layer of square or polygonal stone slabs some 0.5m across and 125mm thick. An alternative paving layer might have been 300mm of gravel concrete. A surface camber of 1:20 from crown to kerb was provided to shed surface water. The mortar used in road construction usually had a lime base, with a sand/limestone ratio between two and three

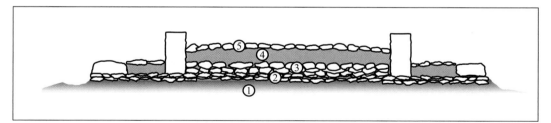

The Romans adapted their road design according to its importance and usage, and availability of local construction materials. They understood fully the importance of layered construction to give a good wearing course, distribute load to the foundation and prevent water reaching the foundation, as shown in this typical section of a major trunk road. Keeping the road surface well above natural ground surface and the provision of side drains were further measures against possible softening of the foundations by water.

1. *pavimentum: well-rammed dry earth*
2. *statumen: 200–300mm thick, flat squared stones*
3. *rudus: 300–500mm, stone blocks in lime mortar*
4. *nucleus: 300mm, concrete with sand, crushed stone, broken tiles or other local materials*
5. *summum dorsum: 200–250mm, cobbles or slabs in mortar*

to one. Where available (as in Statius' description) volcanic ash with the properties of *pozzuolana* was used to make a much stronger mortar, a typical mix consisting of one part lime, three parts *pozzuolana* and up to three and a half parts of sand.

Kerbstone walling, usually placed on top of the third layer, served both to confine the paving layer and define the width of the road pavement. Alternatively, the kerbstone walling was built up from the subgrade, thus acting as a retaining wall for the full depth of the road construction. In some cases the roadway extended for a width of up to 3m beyond the kerb to provide side-paths.

On compact or hard ground, such as in North Africa and Syria, the roads often consisted of simply forming and smoothing the road profile to take wheeled traffic. In contrast, in order to cross marshy areas, they resorted to the ancient corduroy technique of building up the road on a base of criss-cross timbers and/or brushwood. Another form of construction commonly used was to level and compact the natural ground, then cover it with a gravel layer tamped or rolled using wooden or stone rollers. Stones would be gathered from nearby to mark the kerb line, and the drainage ditch would be excavated to about 2m depth. In Britain the roads were commonly built up on broad, gently cambered embankments or 'aggers' sometimes up to 10m or more wide, formed mostly from soil taken from wide, flanking drainage ditches. The road pavement itself sometimes consisted of no more than a single layer of stone, flint or gravel, or if there were nearby workings, waste products such as cinders or iron slag.

While the Romans had no peers as road builders in the ancient world, their prodigious efforts should not detract from the admirable achievements of the Chinese in road construction. It is an interesting parallel between the two civilisations that their main activities in road construction coincided chronologically: the major period in Roman road building lasting from the construction of the Via Appia in 312 BC until about AD 200, while the major period in Chinese road construction spanned the Ch'in and Han Dynasties from 221 BC until AD 220. Needham has estimated that, at the conclusion of the Han period, China had a total length of main roads between 32,000 and 40,000km, compared with a peak of 77,550 km in the Roman system. The land area of the Roman Empire at the time of Emperor Hadrian (AD 117–138) totalled about 4.5 million km², compared with some 4 million km² for the Chinese Empire towards the end of the Han period.

The unification of China under the Ch'in ruler Shih Huang-ti in 221 BC presented conditions favourable for the construction of a major network of roads, both to consolidate the empire militarily and to promote commercial and trading activities. One series of Imperial Highways serving the eastern part of the country radiated from a point close to modern Loyang, the most northerly route crossing the old course of the Yellow River shortly after leaving Loyang, then largely following the course of the river until terminating at a point close to modern Beijing. The most southerly, and longest, route of nearly 1,900km headed due east to Phei, before turning south to cross the Yangtze River near modern Nanking and terminate near Suchow, capital of the former state of Wu. An intermediate road went direct to Lin-Tzu in Shantung Province.

A remarkable series of roads built by the Ch'in and added to, and upgraded, by the Han ran through the high and precipitous passes of the Ch'ing-ling Shan, to open up the country to the south and south-west on Chang-an and to gain access to the rich

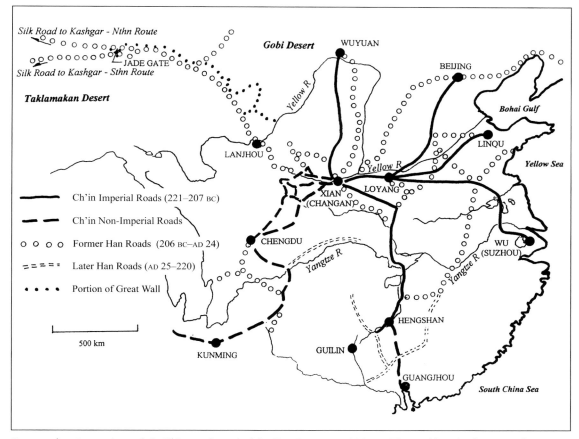

Routes of ancient main roads in China at the end of the Han Dynasty in 220 AD. The total length of main roads in China, between 32,000 and 40,000km, equalled about one-half of the main road system of the Roman Empire. (After Needham et al.)

agricultural land of Szechuan Province. The various roads through the passes eventually linked up to pursue a course, first to modern Chhengtu, then south-east to Chungking, whence to Kweiyang and south-west to Kunming. Extreme measures had to be taken to push the roads through the high mountain passes, including carrying them across raging torrents on high timber trestles (forerunners of those timber railroad structures familiar to devotees of western films), and carrying roads across cliff faces on projecting shelves formed by inserting timber putlogs into holes driven into the rock, a technique also used by the Romans in the Alps and illustrated on Trajan's column.

A highway linking to the Silk Road headed north-west out of Chhang-an, at first negotiating a number of gorges and high passes, until it reached modern Lanchow on the Yellow River. Beyond Lanchow it followed the line of the Great Wall, which protected the road on its northern side through mountainous terrain, before descending, across many streams and alluvial fans, to Suchow on the edge of the Gobi Desert. It then headed into the desert to Ansi, where the road gave way to the two great caravan trails taking northerly and southerly routes around the Taklamakan Desert in the Tarim Basin to meet up again at Kashgar, whence the trail proceeded to Ferghana and Samarkand.

Many lesser roads complemented the Imperial Highways and main roads. Roads were categorised according to their width, generally ranging from one-width to five-width, each width corresponding to about 3m. Tracks and paths suitable only for pedestrians, animals and handcarts were not included in this grading. A one-width road could take a single chariot, a two-width road could take two chariots abreast and so on. Main streets within capital cities sometimes exceeded the maximum five-width of country roads, with a maximum nine-width having been recorded. It is of interest that during Ch'in and Han times the gauge between chariot wheels corresponded very closely to the standard railway gauge of 1.435m (4ft 8½in) today. Needham draws attention to the similarity between the range of road widths in China to those in India, where an ordinary chariot road had a width of 2.5m, a royal chariot road 7.3m and roads to military stations 14.5m. He also draws some comparisons with Inca roads.

The pavements of major Chinese roads appear traditionally to have been made up of a layer or, more likely, layers of rubble, broken stone or gravel, tamped down using metal or wooden rammers to achieve a compact watertight mass similar to water-bound macadam. In an essay criticising the excessive road building of a previous emperor (mostly benefiting, it seems, only the particular emperor and his successors), Chia Shan, a counsellor writing about 178 BC, and cited by Needham, states that: 'The road was made very thick and firm at the edge and tamped with metal rammers.' As users of sealed country roads know well, the edges deteriorate more rapidly than the rest of the pavement and the strengthening of the edges, described by Chia Shan, was a sensible piece of civil engineering learnt by observation and experience. The weakness at the edge of a cambered watertight pavement arises in part, at least, from water being shed from the surface and softening the soil adjacent to the edge of the road, thus reducing the support that it can give to the road. Deepening the pavement at this point, as practised by the Chinese, took it down to a depth where softening of the soil would have been much reduced.

When Hernando Cortés, farmer and soldier, sailed from Cuba to Mexico in 1517, he made his first contacts with a number of settlements around the coast of the Yucatan Peninsula. These were settlements of the Maya people or, more correctly, the remnants of the once great Mayan civilisation, which flourished in Central America between AD 300 and AD 900. Much remains unknown about this civilisation, but they had the technological and managerial know-how to build a wide variety of massive temple structures, which suggests a strong priestly control of their everyday lives. Their intellectual sophistication is attested to by the development and use of an accurate calendar and a partly deciphered form of writing.

At its peak, the Mayan civilisation did not consist of the usual pattern of one or two major cities dominating and controlling the surrounding lands, but rather of ceremonial and administrative centres, with only limited urbanisation, surrounded by essentially rural populations. Traces of an extensive system of roads called *sacboebs*, which linked the centres, can still be seen today. Much of the terrain is extremely rough and the *sacbe* form of construction was developed to cope with this, while maintaining the road surfaces to an alignment and level. The roads consisted of a base layer of coarse limestone layers with variable depth to cope with the uneven ground surface, topped by coarse gravel-size limestone with a surface layer of well-compacted finer-grained limestone, giving a

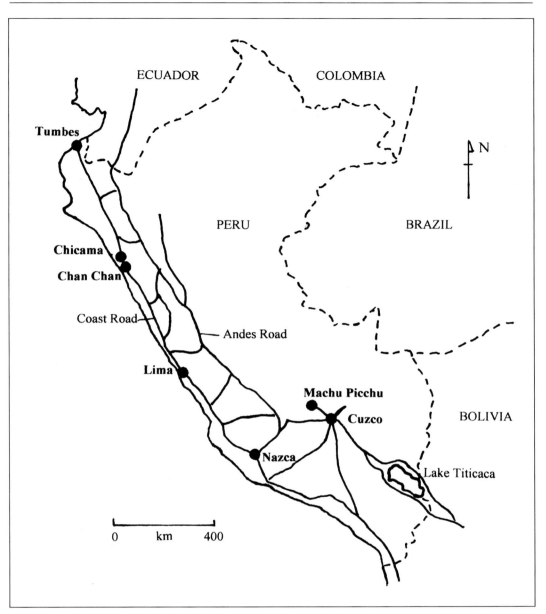

The Inca road system has been likened to a rope ladder, one continuous road following the coast for some 3,500km and a corresponding road through the Andes, with connecting roads and tracks between them.

macadam-type pavement. It has been suggested by geotechnical engineers that a stone cylinder 4m long, 0.65m in diameter and weighing 5 tons, found beside one of these roads, could have been used to compact the road surface, but this view has been challenged by some archaeologists. Tests have shown that a gang of fifteen men could have pushed the cylinder over the road surface. Across some of the deeper depressions the necessary depth of the *sacbe* construction was such that the base layer of limestone boulders became, in effect, a causeway some 8–10m in width.

At the time of the Spanish conquest in 1531 the Inca Empire stretched in a strip down the west coast of South America from near the present Ecuador–Colombia border to south-central Chile, more than 3,500km long and up to 400km wide, with the Royal Road through the Andes Mountains running the whole of this length. A second road from Tumbes, where Pizarro landed with 180 men and 27 horses, followed the coast, crossing the sandy deserts to link up the irrigated valley settlements. Interconnecting roads linked the two major roads, resulting in a network likened by some writers to a rope ladder running the length of the empire.

The Incas ruling the country at the time of the Spanish invasion encapsulated an extended family or clan of thousands of cousins or other distant relatives, who held all the official army and civil service posts throughout the empire, which was nevertheless governed in a well-organised and essentially humane manner. The Great Inca himself was ruler and Chief Priest, held in awe as a god-like figure by his subjects. Shortly before the Spanish arrived, dissension within the ruling family had led to the murder of the Great Inca Huascar by his half-brother Atahualpa, leaving the people in a disoriented and disillusioned state, with little inclination to resist the Spanish as they advanced into the mountains. In making their advance the Spanish made full use of the Inca road system, which had enabled the Great Inca, from his capital at Cuzco, to maintain a tight control over his empire of 900,000 km². Demonstrating bravery and deceit in equal measure, the Spanish seized Atahualpa, whom they thoughtfully baptised before garrotting him. Surviving a half-hearted revolt by the natives, the Spanish assumed control of the Inca Empire, but could not establish full order until 1548 because of bloody infighting amongst the Spanish themselves.

Incan army engineers initially constructed much of the road system, discovered and exploited to his great advantage by Pizarro, when they expanded their empire, from their capital at Cuzco, with astonishing rapidity in the second half of the fifteenth century, until a third of the South American continent came under their control. Without the roads much of the country would have been impassable. Pachacuti, an emperor in the mould of Alexander, set in train these remarkable conquests, by first advancing north to subdue the tribes of the Cordillera Vilcabamba, building roads as he went and hilltop temple-fortresses such as Machu Picchu. He then turned his attention to the area around Lake Titicaca and the deserts to the south of it. But after ten years of campaigning he retired to Cuzco to build it into a great fortress and administrative centre. He left the military actions to his son Topa, who extended the Inca Empire further north to subdue Quito, capital of Ecuador, which became a favourite resort for the Inca royal family. He then attacked the coastal Chimu people, who quickly surrendered.

With the military campaigning over, upgrading of the roads followed, the Inca impressing men from the local villages to do the work, leaving the women to tend the llama flocks and till the fields, producing the staple crops of potato, quinoa (a millet-like grain), maize, pumpkins and other vegetables. They kept a third for themselves and the rest went to the State. Farm produce constituted much of the traffic on the roads, carried by llamas or humans to feed the army, to provide for the temples and the schools, and to

A stepped portion of Inca Road surmounting a steep cliff.

be put into storage against the possibility of a natural disaster. This was an important element in an enlightened social security system established by the Incas.

As the Incas did not have (or need) wheeled vehicles, they made their roads to suit foot traffic, specifically human beings and the sure-footed llama, which they used solely as a pack animal. Most of the roads they left unpaved, but well drained, keeping as far as possible to contours well above river flood levels, and raising the roads on causeways to cross unavoidable low-lying water-logged ground. They made great efforts to maintain road widths of 5–7m, hacking into mountainsides and supporting the road with massive stone retaining walls or tunnelling through geological obstructions. Striving to maintain straight alignments, they climbed steep slopes by constructing stairways paved with stone slabs, and threw suspension bridges across deep ravines. Stone paving was restricted to such stairways, to stretches liable to flooding and to the approaches to important towns.

The roads could only be used for official state business. They served the armed forces, state officials and for the transport of tithes; and they served the Inca's highly efficient courier service, which ensured that messages, even from the remotest ends of the empire, reached Cuzco within seven days. Relay stations, consisting of small circular structures on raised platforms, were provided at 2.5–3km intervals for the couriers, each of whom ran one interval, passed on his message to the next courier, then rested in readiness to run the next interval. As the Inca had no written language, messages had to be relayed orally with the support of memory aids, particularly with respect to numbers and dates, in the form of *quipus'* which were made up of coloured and artistically knotted cords. At intervals of about a day's walk – 20–30km – resthouses provided shelter and sustenance for official travellers and served as supply stations for the military.

5

KEEPING OUT THE ENEMY

The settled way of life adopted by Neolithic communities along the great river systems entailed the taming and herding together of sheep, goats, horses, cattle and other animals, and both these and the villagers needed protection from prowling animals and humans. This protection commonly came from occupying an area of land surrounded by ditches and embankments. Protection was often necessary, too, against floodwaters overtopping riverbanks, and again earthen banks or levees presented the obvious solution. Indeed this remains the case even today.

Early Neolithic enclosures known as 'causewayed camps' date back to about the middle of the fourth century BC in southern Britain. These consist of circular or oval areas of land surrounded by a series of concentric ditches, with the excavated soil thrown up to form embankments along their inner edges, and interrupted at intervals by short, unexcavated, sections or causeways giving access to the site. The exact purpose of these enclosures is conjectural. They may have been multi-functional, providing suitable areas for corralling cattle, for various meetings, for regular markets and annual fairs, and at times of conflict providing some measure of protection for the local populace. The discovery of many arrow-heads in the causewayed camp at Crickley Hill, on the edge of a high limestone scarp in Gloucestershire, certainly suggests it served as a fortified retreat, as does the great depth of the outer ditches at two of the larger sites at Windmill Hill near Avebury and Maiden Castle in Dorset.

The massive earthworks seen at Maiden Castle today immediately mark it as a defensive structure. It is one of up to 3,000 hill forts constructed in Britain from the late Bronze Age through the Iron Age with the arrival of iron-working people from mainland Europe, the earliest dating from about 850 BC. They varied greatly in their origins, size and dispositions. The first earthworks on the Maiden Castle site go back to Neolithic times, when a causewayed camp was constructed at the eastern end of the hill with two concentric rings of steep-sided, flat-bottomed ditches about 1.5m deep. These embraced an area of about 4 hectares. Superficially, none of this is obvious today and the site seems to have passed out of use by 3200 BC, until some centuries later when a bank barrow, about 1.5m high and 546m long – the longest in Britain, and contained between parallel ditches 18m apart – was built along the northern crest of the hill, crossing over the western perimeter of the causewayed camp.

Maiden Castle Iron Age hill fort, built around 350 BC and upgraded in the second century BC, fell to the Romans in AD 43.

Over two millennia elapsed before renewed activity on the site saw the construction, around 350 BC, of an Iron Age hill fort, mostly overlying the Neolithic causewayed camp, and with entrances on the east and west sides. It had a single rampart 3m high and 4m wide, revetted with timber, and a ditch 15m wide and 6m deep. This fort fell into disrepair after about a century, the rampart collapsing into the ditch, but an apparently urgent need for a defensive structure saw it refurbished and extended westwards across the whole hill top and encapsulating 19 hectares in area. Around the middle of the second century BC major works were undertaken to repair and greatly improve the defences, including double ramparts along the north side and treble ramparts along the south, the massive inner rampart having a drop of 15m from its crest to the bottom of the ditch. Quarries within the enclosed area provided limestone blocks to reinforce the back of the rampart. In the first century BC all the defences were upgraded to the standard of the inner rampart and the entrances to the site made much more complex, with, additionally, a sentry box and platform for slingers at the eastern end, and the eastern gateway revetted with stone quarried two miles away. A stash of 22,260 slingstones in a nearby pit came from Chesil Beach 13km away. Despite these massive defences the fort fell to the Roman General Vespasian (later to become emperor) in AD 44, the finding of thirty-eight bodies near the east gate indicating that this may have been the area of fiercest fighting. The site was finally abandoned a few years later in favour of the new Roman settlement of Durnovaria (modern Dorchester) in the valley below.

Jericho has claims to be the oldest, still extant, town in the world. A permanent town occupied the site as early as 8000 BC, protected by a wall 6m high and 800m long, made up of carefully placed, small, unworked stones without mortar, and a stone tower 9m in diameter, with a central staircase, which is still standing to a height of 9m. A lime mortar was used in later walls and wall repairs. An excavated ditch 8m wide and nearly 3m deep, hacked into the solid rock, provided additional security.

While achieving an impressive cultural development in itself, Jericho seems to have had little or no influence in the later development of civilisations in the region, even though a succession of rebuilt walled towns occupied the site, with occasional periods of abandonment, after the destruction of the first town, possibly by fire or earthquake. Whatever grain of truth there may be in the biblical account of the walls falling down, it did not happen in the time of Joshua, in the thirteenth century BC, but at least 1,000 years earlier. According to the Book of Joshua, Chapter 6, the Lord told Joshua to instruct seven priests to take up the Ark of the Covenant and trumpets of ram's horns and, on the seventh day of the siege, carry these seven times around the city, after which they should make a long blast with the trumpets. On hearing this blast, the besieging Israelites should 'shout with a great shout', whereupon the walls would fall flat. And so it came to pass. Supposedly.

Babylon first came to prominence as early as the eighteenth century BC, under the Semitic king, Hammurabi, who brought virtually the whole of Mesopotamia under his control. He set up an administrative system for the whole country and, most notably, had the traditional laws of Mesopotamia codified and inscribed on basalt stelae, one of which, 2.1m high, was discovered in Susa by the French Assyriologist, Jean Vincent Scheil, in 1901. Babylon's dominance lasted only until 1595, when it succumbed first to the Hittites and subsequently to Assyria. In the seventh century BC the Babylonians drove out King Sennacherib of Assyria's appointed King of Babylonia, his own son Ashurnadishum, presumably in the belief that they could withstand any assault Sennacherib could launch against them. But they underestimated the technological and military genius of this great Assyrian king. He sacked the city in 689 BC, laid waste to it, and massacred the people. He then directed water from the Euphrates through the city and left it a wilderness. With this accomplished, he transferred his own capital from Assur to Nineveh, surrounded it with a wall of baked brick interlayered with asphalt, clay and reeds, and turned it into a beautiful city of gardens and canals, copiously supplied with water from sources up to 50km distant.

Babylon's retaliation, when it came, was devastating. Forming an alliance with Scythians and Medes, the Babylonians conquered Nineveh in 612 BC and razed it to the ground; unlike Babylon itself, it was never to rise again. Babylon entered a new golden age under the Chaldean kings, Nabopolassar and his son Nebuchadnezzar, the latter ruling for forty-three years from 605 BC. The city now achieved its greatest size and splendour, protected by an enormous double brick wall, built of baked brick with bituminous mortar and occasional layers of wattled reeds between courses. Entrance to the city was through the famous Ishtar Gate leading to the Processional Way.

Notwithstanding the massive walls, Babylon's second great period lasted only sixty-six years, surrendering peacefully and bloodlessly to the Persian king, Cyrus the Great, in 539 BC. Cyrus ruled the city intelligently and well, respecting its religious traditions, and its trade and commerce; but rebellions by ambitious pretenders to the Babylonian throne, claiming to be sons of its last king, Nabonidus, were put down forcibly by Cyrus' successor, Darius, with attendant damage to the city and the crucifixion of 3,000 of its citizens. The city suffered further damage, including substantial destruction of its walls, rendering them ineffective defensibly, when Xerxes put down another rebellion in 482 BC. In destroying, also, the shrine of Marduk, and carrying off the statue of the god himself, Xerxes reduced Babylon to the status of a provincial town, and it was this reduced state of a once-great city that Herodotus visited. The descriptions of the defensive walls given by Herodotus seem to relate more to the city at the height of its powers than to the walls he actually saw, although they may still have been impressive in their reduced state:

Babylon lies in a wide plain, a vast city in the form of a square, with sides nearly 14 miles long and a circuit of some 56 miles, and, in addition to its enormous size, it surpasses in splendour any city of the known world. It is surrounded by a broad, deep moat full of water, and within the moat there is a wall 50 cubits wide and 200 high (the royal cubit is 3in longer than the ordinary cubit). And now I must describe how the soil dug out to make the moat was used, and the method of building the wall. While the digging was going on, the earth that was shovelled out was formed into bricks, which were baked in ovens as soon as a sufficient number were made; then using hot bitumen for mortar the workmen began by revetting with brick each side of the moat, and then went on to erect the actual wall. In both cases they laid rush mats between every thirty courses of brick. On the top of the wall they constructed, along each edge, a row of one-roomed buildings facing inward, with enough space between for a four-horse chariot to pass. There are a hundred gates in a circuit of the wall, all of bronze with bronze uprights and lintels.

The great wall I have described is the chief armour of the city; but there is a second one within it, hardly less strong, though smaller.

It is hard to reconcile the vast city in the form of a square, as described by Herodotus, with that excavated by the German architect and archaeologist, Robert Koldewey, and his dedicated team of 200, over a period of nineteen years from March 1899. His excavations revealed a triangular area on the eastern side of the Euphrates bounded by a fortified wall (the outer wall) with a total length of about 9km forming two sides of the triangle with the river forming the third. The inner city area, roughly rectangular in shape, with side lengths of about 2.6km by 1.5km, sheltered within this triangular area, and was surrounded by its own wall (the inner wall). Part of the inner city extended across the Euphrates, with a bridge connecting the two parts. Greater Babylon occupied an area of some 850 hectares, and the inner city nearly 400 hectares, with a population estimated at about 80,000. At its height Babylon was the greatest city in the world, but did not occupy a square area with sides 14 miles (22.5km) in length, as described by Herodotus, who was prone to exaggerate such numbers.

The outer, protective wall consisted of a triple construction, comprising an internal wall of sun-dried mud-brick 7m wide and an external wall of kiln-baked brick 7.8m wide, with 12m width of rubble filling between, giving a total width of 26.8m, corresponding very closely to the width of 50 royal cubits claimed by Herodotus, as the royal cubit is slightly more than half a metre. The 7.8m width would have been adequate to accommodate a row of single-roomed buildings on each face, leaving sufficient space between them for a four-horse carriage to pass. Projecting turrets or towers, at regular intervals of about 50m, served to strengthen the outer wall, and five fortified gates, covered in bronze, provided access through it. A brick-lined moat running the length of the wall, fed from the Euphrates, provided not only additional protection from external aggressors, but also, according to Herodotus, the earth used to make the bricks for the wall.

The claimed height of 200 cubits (over 100m) is clearly excessive and approximate heights may have been 25m, 15m and 20m respectively for the internal wall, rubble and external wall, which accord well with a wall height of 50 cubits, and tower heights of 60 cubits given by Strabo. Herodotus' exaggeration of the height hardly matches that of Nebuchadnezzar:

Remains of the Ishtar Gate at Babylon excavated by Koldewey. This main entrance led directly to the Processional Way, which ran through the centre of the Inner City. (From Hammerton)

I completed the building of the sublime city of Babylon. My father had encircled the city by walls of asphalt and brick. I raised a third wall beside the others of asphalt and brick and joined it to my father's walls. I set their foundations on the very bosom of the underworld and raised their summit as high as a mountain . . .

The wall surrounding the inner city area also consisted of a triple construction, comprising an internal wall of mud-brick 6.5m in width and an external wall, also of mud-brick, 3.7m wide, the two separated by a 7.2m-wide space filled with earth or rubble. Turrets strengthened the internal and external walls at regular intervals of 18m and 20.5m respectively. A wide moat lined with baked bricks ran the length of the wall. Eight gates provided access through the inner city wall, with the most important, the spectacular arched opening of the Ishtar Gate, giving direct access to the Processional Way, which ran through the centre of the city, parallel to the Euphrates River.

After endlessly digging through drab brick, rubble and mud, it must have been a spellbinding moment when Koldewey's excavators uncovered the brilliant blue-glazed fired bricks of the Ishtar Gate, surprisingly not mentioned by Herodotus. But there survived only very few of the several hundred bricks which had been moulded with partial reliefs in such a way that, when placed in position, they revealed brilliant depictions of bulls and the Sirrush, the latter a fabulous monster with a scaly body, serpentine neck and tail, and a

A reconstruction of the Ishtar Gate.

The Sirrush was a fabulous creature with a scaly body, a serpentine tail, neck and head, a forked tongue and horns. Its front feet were those of a large cat and its back feet those of a griffin or a raptor. (From Koldewey)

horned head with a protruding forked tongue. The front feet are like those of a great cat and the hind feet are griffon or eagle claws. These figures adorned not only the glazed brick facing of the gate structure above ground, rising to some 23m, but also the unglazed portion, sunk through 15m of sand below pavement level. This was the portion revealed by Koldewey's excavations.

Nebuchadnezzar's boast that he set the foundations of his walls 'on the very bosom of the underworld', and thus at great depth, was meant to impress, but also had a practical significance. Constructing heavy structures such as walls, temples and palaces on the soft alluvial deposits of the Tigris and Euphrates Rivers must have presented Mesopotamian builders with a constant problem of settlement. It could only be overcome, in part at least, by putting the foundations at great depth. Internal joints were built into the structure of the Ishtar Gate to reduce the effects of settlements further, which otherwise would have caused unsightly cracking of the brittle, glazed brick wall.

The massive reconstruction programme for Babylon pursued by Nabopolassar and his son Nebuchadnezzar would have required mass production of sun-dried mud-brick, the principal and, indeed, the only readily available building material. Both mud and sun were plentiful. Mud dug from the moats would certainly have been used to make bricks for the construction of the walls as stated by Herodotus, but would have provided only a portion of the total amount required. The bricks were formed by pressing the mud into moulds, commonly 320–350mm square and 115mm thick, and then sun-dried and stacked.

Herodotus clearly had first-hand knowledge of Babylonian building techniques and may have witnessed some building activity during his visit there. He makes particular mention of the use of bitumen, and is correct in saying that it came from near a town called Is (Hit), where it occurred in lumps in a small tributary of the Euphrates from where it could be easily shipped to Babylon. Bitumen occurs naturally in a number of areas of the Middle East as a result of upward seepage through fissures in the ground. In order to be used as mortar the bitumen had to be heated, as noted by Herodotus, so that it could be spread and would adhere to the bricks. He also mentions the insertion of rush matting between every thirty courses of brick. The purpose of such matting is to lend tensile strength to the structure, which would mitigate the effects of shrinkage or settlement; but to have been effective it would need to have been more closely spaced than every thirty courses of brick. According to Koldewey, the mortar between the brick courses consisted of a layer of asphalt with an overlying layer of clay, with the latter replaced by a layer of reed matting in every fifth course. In order to avoid staining, bitumen was used more sparingly in walls of glazed brick and, towards the end of Nebuchadnezzar's reign, bitumen was being generally replaced by an inferior lime mortar with varying amounts of bitumen added to it.

Although used relatively sparingly, baked bricks, such as those used in facing the Ishtar Gate structure, would have consumed in their firing large amounts of fuel, a scarce commodity in Mesopotamia. Reeds and palms were no doubt readily available, as was dried animal dung, which is still used as a domestic fuel in some countries today; but it would have been difficult to attain the high temperature required using these materials. Bitumen may have been used, as may timber shipped down the Euphrates from mountains to the north. As the animal reliefs extended across some forty bricks, a corresponding number of different moulds or mould facings would have been required to shape the bricks before firing.

The first phase of Sumerian dominance in Mesopotamia ended in about 2350 BC, when the walls of these city-states proved inadequate against the attacks from their northern Akkadian neighbours under King Sargon I, whose capital Akkad, believed to be near Babylon, has never been identified. Sargon's conquest brought a degree of unity to the city-states of Sumer that had hitherto been lacking. The Akkadian language, based in its written form on Sumerian cuneiform, gradually supplanted the latter and became the lingua franca for the whole Near East area for the next two millennia.

The Sumerian city-state of Ur has achieved particular prominence, largely as a result of the remarkable findings revealed by the excavations between 1922 and 1934, directed by Sir Leonard Woolley, including a remarkable royal grave and evidence of a flood pre-dating the biblical version by more than 2,000 years. Ur underwent a renaissance under King Ur-Nammu, of the Third Dynasty, and became the capital of Sumer. Although the site is now 200km inland, it occupied, at that time, a prime position on the now-retreated Persian Gulf. Canals linked it directly to the Gulf waters, allowing ships to berth in harbours at the very edge of the city. It was now a city roughly oval in shape, 1.2km long and 800m wide, surrounded by a mud-brick wall 8m high, 23m wide at its base and 4km long.

Following its renaissance under Ur-Nammu, Ur went into a long-term gradual decline, until it underwent a further building revival under the direction of Nebuchadnezzar and his son Nabonidas. Nebuchadnezzar's principal contribution was the construction within the city of a great mud-brick wall surrounding the temenos, or sacred area dedicated to Nannar the Moon-god, roughly rectangular in shape, some 350m by 200m. Woolley found well-preserved remnants of the wall of up to 2m in height. Thought to have been about 9m high when constructed, the wall was 10m wide and consisted of inner and outer walls separated by chambers, the flat tops of which formed a broad passage along the top of the wall, which could be trafficked by defending troops. The external face was decorated with double vertical grooves. Six fortified gateways gave entrance to the sacred area, the main one, distinguished by a high tower, leading directly to the entrance of the Great Courtyard in front of the ziggurat. Some puzzling changes in the direction of the wall and differences in foundation levels of adjoining sections are thought by Woolley to have been a result of poor collaboration between various gangs, each gang having its own section.

High masonry or brick walls remained the most effective means of protection for cities and urban settlements until the invention of cannon. It was the breaching of the walls of Constantinople by the Ottoman force of Sultan Mehmet that exposed the weaknesses of such walls to cannon fire. Previous to this, any assault on high defensive walls manned by determined defenders constituted a hazardous exercise whether using scaling ladders, siege towers or tunnelling to undermine the wall or gain access below it. High defensive walls are, consequently, the most common feature linking changing civilisations over many thousands of years and different peoples in many lands.

Some idea of the means of besieging high defensive walls is afforded by a drawing, dating from about 1280 BC, on the south wall of the Great Hypostyle hall of the Ramesseum in Egypt: it depicts two lines of turreted curtain walls protecting the Hittite city of Dapour while under siege from Egyptians, who are climbing scaling ladders, against defenders armed with bows and arrows and lances. An even more graphic Assyrian illustration in bas-relief, dating from about 800 BC and now in the British Museum, shows the attack on a city by Ashurnasirpal using a combined tower and battering ram. Weapons include swords, bows and arrows and stones thrown by hand. As well as employing battering rams to destroy the walls, the men are shown working separately in pairs undermining the walls.

The rocky Anatolian hilltop site of Hattusas, with its natural protections of deep gorges, rocky pinnacles and steeply sloping terrain with plunging cataracts and precipices on two sides, was first populated in the seventeenth century BC and further protected by fortified walling. By the time the Hittite civilisation reached its peak in the thirteenth century BC this had become a massive double wall 4km in length. The wall consisted of a cellular structure of masonry with earth filling, supporting a sun-dried mud-brick superstructure. Several massive gates gave access to the site, including the King's Gate, Lion Gate and Sphinx Gate, the first two named after reliefs adorning the doorways and the last by pairs of sphinx-like sculptures protecting each of the doorways. Each gate consisted of inner and outer doorways, separated by an enclosed chamber, and were flanked by huge curved stone uprights. Adjacent to these were fitted masonry stone blocks of cyclopean proportions which added to the overall massiveness of the whole structure.

Entering the Lion Gate at Hattusas must have been a daunting experience for uninvited visitors.

Discovery of the Hittites and their capital Hattusas came relatively late in the history of archaeological exploration in western Asia and caused some excitement when some of the 20,000 tablets found in 1907, and deciphered in 1915, showed that they spoke an Indo-European tongue. Unlike Mesopotamia the Hittites possessed rich sources of metallic ores, particularly copper, iron and silver, and consequently became skilled metallurgists. Their knowledge of the working of iron, and their possession of it, may, for a short time, have given them some edge in warfare. More importantly, it put them in a strong, even monopolistic, trading position in this much sought after and prestigious commodity. Other than in metallurgy they were not notably innovative, and absorbed much of their culture, including the cuneiform script, from Mesopotamia. In 1280 BC, at Kadesh, they held off a great Egyptian army led by Ramses II, but their final demise came around 1185 BC at the hands of invaders from the west, known for want of a better or more precise term as the Sea Peoples. Assyria became the dominant power in the region.

An alliance of Babylonians and Medes destroyed the Assyrian Empire in 612 BC and established the two allies as the major powers in the region. The Medes had in fact thrown off the yoke of Assyria a century previously and united under a king Herodotus names as

Deioces, a village elder known for his integrity and fair dealing. He established the capital of the now-unified tribes at Ecbatana (modern Hamadan in Iran), which vied in importance with Babylon and, judging by Herodotus' description, was hardly less impressive, at least in its fortifications:

> Once firmly on the throne, Deioces put pressure on the Medes to build a single great city to which, as capital of the country, all other towns were to be held of secondary importance. Again they complied, and the city now known as Ecbatana was built, a place of great size and strength fortified by concentric walls, these so planned that each successive circle was higher than the one below it by the height of the battlements. The fact that it was built on a hill helped to bring about this effect, but still more was done by deliberate contrivance. The circles are seven in number, and the innermost contains the royal palace and treasury. The circuit of the outer wall is much the same extent as at Athens. The battlements of the five outer rings are painted in different colours, the first white, the second black, the third crimson, the fourth blue, the fifth orange; the battlements of the inner rings are plated with sliver and gold respectively. These fortifications were to protect the king and his palace; the people had to build their houses outside the circuit of the walls.

About a hundred years before the time of Herodotus, the Medes had been overcome by their Iranian (Aryan) kinsmen, the Persians, who proceeded to establish an empire comparable to the later empires of Macedonia and Rome. Under Persian domination Ecbatana remained an important centre and, because of its high elevation – 1,830m above sea level – it became the summer capital of the Persian Empire. As such it was an important centre of communications, and a special spur road connected it to the Royal Road through Persia.

After the routing of the Persian navy at Salamis in 480 BC and the subsequent defeat, at Platea in 479 BC, of the Persian army, the Athenian leader, Thermistocles, who had been responsible for building up the Athenian navy, now set about strengthening the fortifications of Athens by rebuilding and heightening its defensive walls. His allies, the Spartans, who had led the Greek armies, opposed this work, claiming it could help the enemy should they return. Thermistocles visited Sparta to reassure and placate the Peloponnesians, but ordered the work to continue apace during his absence, using any available stone, including old buildings, statue bases and even gravestones. He invited a deputation from Sparta to visit Athens to assure themselves that rumours they heard about the work proceeding were not true, but sent word back to Athens that the deputation should be 'politely' detained until his own safe arrival back home. Thermistocles then confessed all to the Lacedaemonians, who had little choice but to accept it in good grace.

Because Piraeus was strategically more important than Athens, Thermistocles also ordered the building of stone walls up to 18m in height around the Munychian Peninsula to enclose the port. As relationships with Sparta deteriorated, the Athenians decided to form Athens and Piraeus into a double town surrounded by a continuous fortification. To achieve this, two continuous diverging walls – the Long Walls – were built over the next twenty years. The northern wall was 8km in length, linking the walls of Athens to the

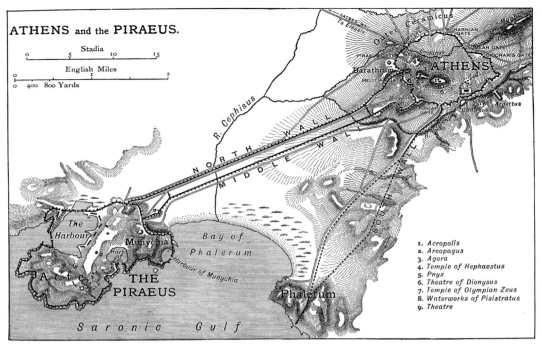

ATHENS and the PIRAEUS.

1. Acropolis
2. Areopagus
3. Agora
4. Temple of Hephaestus
5. Pnyx
6. Theatre of Dionysus
7. Temple of Olympian Zeus
8. Waterworks of Pisistratus
9. Theatre

The ancient walls of Athens and Piraeus. The route of the South Wall to Phaleron is known only approximately. (From Bury)

Piraeus walls near the harbour, while the southern one was 5km in length and linked the walls of Athens to the Phalerum Peninsula. The Bay of Phalerum, which separated Piraeus from Phalerum, represented a weakness in this system of fortification, as an enemy could have landed there at night. Consequently, under Pericles, a third structure, the Middle Wall, was constructed, running parallel and close to the northern wall, linking the Piraeus walls to those of Athens. Herodotus must have been very familiar with these walls. In the second Peloponnesian War, Sparta, with financial and military backing from Persia and led by the outstanding general Lysander, destroyed the Athenian fleet and army, leading to the surrender of Athens in 404 BC. Lysander immediately ordered the destruction of the Long Walls and Piraeus fortifications, which were demolished to the music of flute players. A decade later the Persians, having fallen out with the Spartans and ravished their lands, gave their support to Athens and allowed the walls to be rebuilt. The Greeks were no mean builders of defensive walls and, around 120 BC, a well-known early treatise on military architecture was written by Philon of Byzantium, who had spent time in both Rhodes and Alexandria becoming well versed in the practices of the time.

Homer and Schlieman between them have made the walls of Troy among the most famous in the ancient world. The small hill of Hisarlik, guarding the mouth of the Dardenelles, played host to nine distinct settlements, now designated Troy 1 to Troy 9, over a period extending from before 3000 BC to Roman times, accumulating, in the process, 15m of debris as each settlement levelled that of the previous one and built on top of it, surrounding itself with new mud-brick walls. The city reached its greatest extent, some 200m by 160m, at the time of Homer's Troy, known as Troy 6, when massive walls of limestone replaced those of brick. The walls rose with a steep external batter to a height of 6m, with a 2m-high vertical parapet wall topping this. Building it in straight sections, some 5–9m long, accommodated the curvature of the wall, each section being offset by about 150mm where it met its neighbour. The reason for the offsets is not clear, but Toy notes that the vertical shadows of the offsets impart a great feeling of strength to the wall. Troy 6 was destroyed around 1250 BC, perhaps as a last fling of Mycenaean aggression and perhaps already having been damaged by a strong earthquake. Did Agamemnon, 'King of Men', accompanied by great princes of Greece, really take on the siege of the almost impregnable Troy to rescue the fair maiden Helen, as Homer would have it? Or was the Trojan War started for more prosaic reasons, such as the crucial position of Troy at the mouth of the Dardenelles or even, as has been suggested, a dispute over fishing rights or concern over competition in the textile trade? Perhaps. But why spoil a good story?

As builders of mighty walls themselves, the Mycenaeans must have known all too well the unenviable task they had set themselves in besieging Troy. They surrounded their own hilltop citadel of Mycenae, which dated from about 1500 BC, with cyclopean masonry walls 5–8m thick and up to 12m high which followed the natural contours of the rock. These walls must have been a great comfort to the farmers tilling the fields on the lower slopes, who would have known they had such an impregnable fortress into which they could, if necessary, retreat. Anyone wishing to enter the Acropolis of Mycenae had to negotiate a long ramp flanked on either side by high stone walls and then pass through the

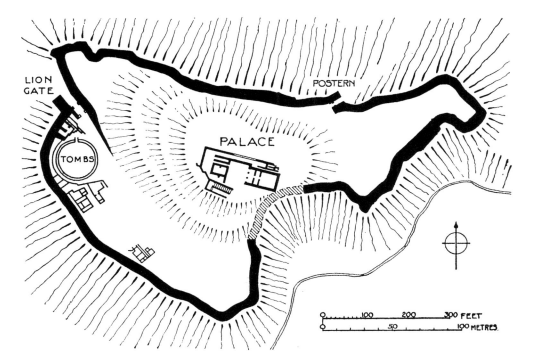

Plan of the citadel of Mycenae. (After Toy)

115

The Lion Gate, Mycenae.

forbidding Lion Gate, defended by a strong tower on its right-hand side. An impressive and dignified approach for a welcome visiting dignitary, or a successful Agamemnon returning from the wars, but a dismaying one for potential besiegers.

Other citadels lost little by comparison, notably Tiryns, with walls reaching as much as 17m in thickness and 20m in height. Later Greeks could not believe that mere humans raised the enormous blocks used in constructing the walls of these citadels, and concluded that the work had been done by one-eyed giants, the Cyclops, making the name synonymous with this type of construction. In most of the Mycenaeans' defensive wall construction, which averaged about 4.6m in thickness, huge cyclopean blocks, at best only partly dressed and often not dressed at all, with small stones or clay filling the gaps, formed the inner and outer faces, separated by a core of rubble or earth. Where appropriate, however, such as in the surrounds of the Lion Gate, to give it added dignity, they used large symmetrical blocks carefully dressed by hammer and cut by saw.

The Mycenaean civilisation, which built the fortresses of Mycenae and Tiryns, disappeared into history around the end of the twelfth century BC, perhaps as a result of

Dorian invasions, or simply as an inability to cope with climate changes and crop failures. Herodotus makes no mention of them, despite the fact that he must surely have been aware of their physical remains.

The Roman author Vitruvius, writing in the first century BC, also made recommendations on the construction of city walls, stressing the need for good foundations and the avoidance of sharp angles which would give assailants protection. Towers should be round or polygonal and not square, as this shape was more susceptible to pounding by battering rams. Various materials, according to availability, could be pressed into service for wall construction – dimensioned stone, flint, rubble, burnt or sun-dried bricks – 'use them as you find them'.

When Constantine had the town of Byzantium, on the Bosphorus, enlarged, between AD 324 and AD 330, to create a city worthy of bearing his name, he destroyed earlier fortifications and built a new system of enclosing walls. After his death in AD 337 the city continued its rapid cosmopolitan growth, and, by the time of the division of the Roman Empire, Constantine's walls no longer enclosed sufficient territory, many houses having already been built on the unprotected land outside the gates. Added to this were increasing threats from beyond: the Huns had crossed the Danube and settled in ostensibly Byzantine territory, and Alaric, who had fought for the Byzantine emperor, Theodosius I, in AD 394, rebelled after that emperor's death in AD 395. Leading the Visigoths, Alaric attacked and sacked Rome in AD 510. In the meantime, Theodosius II, grandson of Theodosius I, had become Eastern Emperor and he commissioned his prefect Anthemius to build land walls around Constantinople some 2km outside existing defences.

The corridor between the 'mighty walls' leading to the apartments of the megaron (principal hall) of Tiryns.

The Theodosian walls protected Constantinople for 1,000 years until breached by Ottoman cannon fire in 1452.

Anthemius built a wall 5.63km in length, 4.5m thick and nearly 12m high with ninety-six projecting battlemented towers up to 189m high; some of these were square and some octagonal in shape. Parts of the walls, and sixty-seven of the towers, suffered extensive damage from a severe earthquake in AD 447, and Theodosius II, now nearing the end of his long reign as emperor (he died in AD 450 after falling from his horse), commissioned prefect Constantine Cyrus to restore and strengthen the fortifications; the threat from Attila had reached its height at this time, his Huns having just taken the Danubian town of Marcianopolis just to the north of Constantinople. As well as restoring and strengthening the damaged wall, Constantine added a lower outer wall, 2m in thickness and with ninety-two towers, hexagonal and octagonal, located at intermediate points between the towers of the inner wall. Beyond the outer wall he excavated a moat 15.2m wide and 5.7m deep.

The Theodosian walls were constructed with a concrete core, faced with limestone blocks, bonded, at intervals of height, with horizontal brick-lacing courses five bricks thick, giving a distinctive pattern that became common in Turkish architecture. These walls kept Constantinople intact and secure against many attacks and assaults for a period of 1,000 years. Avars and Slavs besieged the city, as did Arabs, Bulgars, Russians, Uzes and Patzinats, all to no avail. The fall of Constantinople to the Fourth Crusade in 1204 came about not through breaching of the walls, but through a process of disintegration within the city: social, economic and political. Like much of the rest of Europe the city suffered the ravages of the Black Death in 1347. But the walls remained intact until 29 May 1453, when the Ottoman, Sultan Mehmet, after a siege lasting several weeks, breached the walls with the biggest cannon the world had yet seen.

The Romans built walls not only to protect individual cities, but also to serve as frontier boundaries for their far-flung empire. A famous example was built by the order of Emperor Hadrian around AD 122 to guard the northern frontier of Roman Britain. Constructed with a rubble core bonded by clay or lime mortar, and dressed with stone facings, it extended for 117km, standing typically about 4.5m high and 1.8–3.0m in width. Associated with the wall were ditches on both the north and south sides. The original purpose of the wall may have been in part to mark the boundary and to deter smuggling, cattle raiding and other minor incursions, but it also proved to be an effective barrier, for a time at least, against the northern barbarians.

As well as having to traverse variable terrain, the wall crossed three major rivers, requiring multiple span bridges with stone piers and superstructures of stone arches or timber. At intervals of about 1 mile, small fortlets, about 18m square, were built on the southern side of the wall and contiguous with it. These 'mile castles' controlled crossing points through the wall. Intermediate between these were two towers or turrets at approximately one-third-mile intervals. Small differences in details found in the construction of the walls, and appurtenant structures along different parts of its length, point to three separate legions having been engaged on the work.

All other ancient defensive walls are dwarfed in magnitude by the Great Wall of China. It is not a single wall but many walls, built over a period of more than 2,000 years, from the time of the Warring States (around 700 BC), culminating with the Ming Wall, begun in

Hadrian's Wall crossing the crags towards Housesteads Fort, looking east.

Housesteads Fort on Hadrian's Wall, the north-east barracks blocks.

AD 1368 and completed to a length of 7,300km over a period of some 200 years. Various spurs and bifurcations, built to satisfy local defence needs, have left a bewildering pattern of walls, traversing every conceivable type of topography – mountains, cliffs, deserts, loess regions, river valleys and alluvial plains – stretching from the Yalu River in the east to Turkestan in the west, a distance of 2,700km. In total, some 50,000km of wall may have been built, much of which has long since disappeared, or is little more than a slight ridge or a line of broken stones crossing the landscape.

Prior to 221 BC the rival warring principalities in northern China built beacon towers and blockhouses, which they gradually connected together by walls to suit their own defence needs against neighbouring states. One of these states, Ch'in, defeated and annexed its neighbours, and in 221 BC Emperor Qin Shi Huang established a unified China. He had many of the existing walls torn down and built the first Great Wall, incorporating within it some of the surviving previous walls. The purpose of this Great Wall, which wound its way for a length of some 5,000km, was to protect the feudal, agriculturally rich, China from northern invaders. The Ch'in Dynasty survived for only fifteen years before succumbing to a peasants' revolt, arising in part from the heavy taxation and harsh working conditions imposed by the rulers to build the wall.

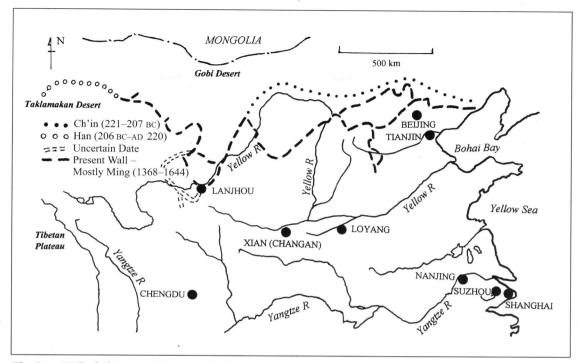

The Great Wall of China, stretching 2,700km from east to west, in fact consists of many walls built over a period of 2,000 years from about 700 BC. (After Needham et al.)

The succeeding Han Dynasty (206 BC–AD 220) undertook an even more ambitious programme of wall building, eventually completing a total length of some 10,000km, in addition to repairing lengths of the existing Ch'in wall. The western section of the Han wall safeguarded the Silk Road, which skirted along its southern side, and thus played a major part in developing Chinese economy and trade with various European and Asian countries. Ruined beacons and tombs along its length have yielded up abundant evidence of this trade – wooden strips and writing on silk, seals, silk and hemp ropes. Wall construction declined after the end of the Han dynasty, but did not cease. Some 6,000km were constructed, together with strengthening of existing walls, up to the time of the conquest of the Yuan (Mongol) dynasty by the Ming in AD 1368. Throughout the 200 years of their rule the Ming worked continuously on building the Great Wall, achieving a length of some 6,500km.

Before the Ming period the builders used the materials most readily available near the site: stone walls over the mountains and earthen walls or embankments across the plains. Where they used earth they fully understood the need to compact it, and the layering of soil from compacting the successive layers can still be seen in surviving fragments of ancient walls. In desert areas the Han builders constructed their walls, up to 5–6m high, of sand and gravel layers, each about 200mm thick, separated by layers of interwoven reeds or tamarisk twigs about 50mm thick, anticipating by 2,000 years the modern technique of earth reinforcement. Great difficulty must have been experienced where only highly

Local materials were utilised resourcefully in the construction of the Great Wall, including sands and gravels stabilised by the insertion of horizontal mats of reeds or tamerisk twigs, as shown here in this section of wall dating from the Han period (206 BC–AD 220).

erodible loess soil was available – sections of wall or embankment of this material facing strong prevailing winds or subjected to heavy rains would have required constant maintenance.

The walls built during the Ming period were much more substantial than their forerunners; they were commonly 7–8m high, and 6–7m wide at the base tapering to 5m at the top. They were made up of inner and outer facing walls 1.5m thick of dressed rectangular stone blocks, typically about 1.5m by 0.6m by 0.5m, or baked bricks made from carefully selected clay, built up on solid granite block foundations and the space between the facing walls filled with compacted earth or rubble. This was topped by a paving of stone slabs or bricks, formed into steps along steeply inclined lengths of wall. Watchtowers of brick or stone, or a combination of these, were built at intervals of about 220m, averaging 11m square and 13.5m high. Transportation of heavy stone blocks, bricks and lime for mortar over considerable distances, and at times up precipitous slopes, represented a profound feat, with human and animal power supplemented only by simple devices such as handcarts, rolling logs, levers, windlasses and hand-operated cableways. Animal power came mainly from sheep and donkeys, their popularity deriving largely from their mountain-climbing abilities.

During periods of peak construction activity, up to two million people may well have been engaged on wall construction. They required food, water and accommodation. Emigrant peasants opened up large tracts of land along the route of the wall in order to supply the food, and conservancy projects were undertaken to supply water for domestic needs and for irrigation. Troops, garrisoned along the lengths under construction, protected the workers from the enemies the walls were intended to keep out and may also have been used to assist in some of the work.

Occupying a site on the southern escarpment of the 1,200–1,500m-high plateau forming the watershed between the Zambezi River to the north and the Limpopo River to the south, the Great Zimbabwe site in southern Africa offered generous resources and undoubted blessings to the people living within its walls: a good, reliable rainfall, freedom from tsetse flies and malaria, woodlands providing ample timber for building and fuel, light soils suitable for crops, open grasslands ideal for cattle, and minerals, notably gold. Its 18,000 or so inhabitants, ancestors of today's Shona people, enjoyed, in the thirteenth to fifteenth centuries, a thriving cattle-based internal economy and a prosperous external trade in luxury goods. Unfortunately, early destructive digging at the site has obscured the full range of economic activity, but beads and other glass objects, and pottery from places

A section of wall composed of compacted clayey earth dating from the Ch'in period (221–206 BC).

The best preserved sections of the Great Wall are the stone lengths, originally totalling 6,500km in length, built during the Ming Period (AD 1368–1644). Up to 7.5m high to the base of the parapets, the walls had dressed stone facings with rubble infill, giving a total base width up to 7.5m. Watchtowers up to 12m high were built at intervals of about 220m along its length. Drainage outlets can be seen projecting from the base of the parapets.

as distant as China, Persia and other near Eastern counties, have been found, and gold was probably a major export. Great Zimbabwe may also have been a hub through which goods reaching the coast of Africa were exchanged for gold and copper objects, and other products, deriving from interior regions of the country.

Stretching about 800m from north to south, the Great Zimbabwe site can be divided into three distinct areas. The northernmost group of structures occupy a *kopje*, or ridge of granite, dubbed 'the Acropolis' by early European visitors. The southernmost, and most impressive, remains are those of the Great Enclosure or Elliptical Building, occupying a flat-topped granite shelf on the other side of the separating valley. The ruins of a number of other smaller structures occur on the shallow slopes of the valley.

The Elliptical Building is by far the largest and most recognisable structure and, indeed, is the largest ancient structure in southern Africa. Its substantial perimeter wall, roughly elliptical in plan, has a circumference of over 240m, with a maximum length of 89m and breadth of 67m, a height of up to 10m and a width ranging from 1.2m to 5m, tapering outwards from top to base. There are entrances to the site at the northern end as well as in the north-east and the west. Notable features within the site include a roughly circular enclosure 21m in diameter, and on the eastern side a secondary wall paralleling the outer wall and close to it, forming a narrow passage between the two. At the southern end of the passage is the most prominent feature within the site, a stone tower of truncated conical form 10m in height and 5.5m diameter at its base, with a second, much smaller, tower nearby. There is no evidence of military threat to Great Zimbabwe at its height, so the question arises why the need for these massive walls; but perhaps when the site was first settled possible threats were envisaged and these pioneer inhabitants might even have been displaced from another area by military activity.

The builders had abundant availability of stone for the construction of the walls in the form of the local granite rock, which forms the bare rounded hills around the site and on which some of the walls are founded directly. This rock has the property of exfoliation, by which surface layers of consistent thickness, usually 75–175mm, rather like onion skins, break away easily, either naturally or with a little human assistance, from rounded masses

The Elliptical Building or Great Enclosure, built from local granite, is the southernmost of the Great Zimbabwe structures. Features within the site include various enclosures, the largest 21m in diameter, and a narrow passage formed by a second wall just inside, and paralleling, the eastern wall. The southern end is occupied by a stone tower 5.5m diameter at its base and originally 10m high. At its height in the thirteenth to fifteenth centuries Great Zimbabwe was the centre of a thriving cattle-based internal economy, and also had lucrative trade links with countries as distant as China.

of rock, as a result of fluctuating day/night temperatures. If necessary the rock can be heated by lighting fires on it and quenching it with water to separate layers from the parent rock. The cracking patterns in the rock allow brick-shaped pieces of rock to be produced, and, as seen at Zimbabwe, most of the walling has courses of stone, each course of uniform thickness resembling brickwork, except for the variation in lengths of the stone pieces, giving an irregular pattern of vertical joints and the occasional coincidence of vertical joints through two or more courses. The taper in the walls was produced by setting each course back slightly from that below it. The regular courses of stone form the inner and outer visible faces of the walls, with the space between loosely filled with less regularly shaped stones. However, the quality of construction varied considerably and in some sections the outer stones are irregular in thickness and poorly coursed or barely coursed at all.

The builders at Zimbabwe also made use of puddled clayey soil with fine gravel aggregate, known as *daga*, which set hard on drying under the sun. It was used as a wall plaster, as well as for flooring and for the walls of dwellings.

At much the same time as Mehmet breached the walls of Constantinople, and the civilisation of Great Zimbabwe had started to decline, the Incas were rapidly extending and consolidating their hegemony over a large area of South America, comprising much of modern Ecuador, Bolivia, Peru and northern Chile. When Pachacuti became the Eighth

The Sacshuaman fortress, up to 20m in height, stretched for 400m across a hillside overlooking Cuzco. Some stones weigh up to 100 tonnes.

The famous 12-sided fitted stone, about 1m high, in a wall at Cuzco, the Inca capital.

Emperor of the Incas in 1438 he initiated a series of military campaigns that resulted in a tenfold enlargement of the Inca Empire. After ten years in the field he retired to Cuzco, to build it into a great citadel and administrative centre, leaving his equally able son Topa to carry on with the military conquests.

The great fortress of Sacsahuaman commissioned by Pachacuti to be built into the steep hillside overlooking Cuzco still presents an imposing site today, despite extensive quarrying by the early Spanish settlers using Indians as forced labour. Only the huge size of its blocks saved it from complete extinction. Three sides of this great fortress comprised triple lines of terraced cyclopean masonry walls up to 20m in height and 400m in length, while the remaining side was guarded by a sheer natural cliff. Within these walls the Inca army maintained its main arsenal and parade ground, sharing the vast area with a temple, a palace for the emperor, a bewildering array of buildings and a reservoir. Re-entrants and salients gave the walls, in plan, a saw-tooth configuration to increase their effectiveness to resist attack.

Masonry walls of perfectly matched polygonal blocks characterised Inca masonry, and substantial remains of such walls can still be seen at Sacsahuaman. The largest blocks have main dimensions up to 5m and weigh over 100 tons. While some of the masonry blocks used came from nearby limestone and diorite quarries, other blocks of andesite had to be transported to the site from up to 20km away over rough terrain with steep slopes. According to Garcilasco, the son of an Indian princess and a Spanish conquistador, writing in Spain in the second half of sixteenth century, it required 20,000 Indians hauling on large cables to move the larger blocks. On one occasion he claimed a block broke loose from its cables, rolling back down the hill and killing 3,000 Indians.

It is impossible, even today, not to be amazed at the remarkable achievement of the Inca masons in fashioning huge shapeless stone blocks, using only stone tools, into polygons with up to twelve straight edges viewed two-dimensionally (up to thirty flat faces altogether) and then fitting these together with the exactness of a jigsaw puzzle. Such precise matching of adjacent blocks would have required first shaping one or both as accurately as possible, then positioning one against the other and marking the high spots: the blocks would then have been separated and the high spots removed by chipping and/or grinding, then bringing the blocks into contact again and repeating the operation as often as necessary to get a perfect fit. As some blocks weighed up to 100 tons this must have been an exhausting exercise. In order to place the stones in the wall they may have been rolled up earthen ramps, a method the seventeenth-century Jesuit priest, Father Cobo, records having seen used in the construction of Cuzco Cathedral: Spanish in design but built by Indians using their traditional methods.

Unfortunately, the fortress of Sacsahuaman had not quite reached completion when Pizarro marched into Cuzco on 15 November 1533. It did not really matter. Pizarro had already captured and garrotted Atahualpa, who, while a prisoner of the Spaniards, had arranged the murder of the only other real threat to the Spaniards, his own brother and rival Huascar. With both Atahualpa and Huascar dead, the Spaniards, making maximum use of their horses and superior weapons, inflicted one defeat after another on the increasingly dispirited Indians, as they advanced towards Cuzco. No fortress at Cuzco, complete or otherwise, could have saved them.

6

ZIGGURATS, RITUAL MOUNDS AND AN ANCIENT ENVIRONMENTAL ICON

As the Tigris and Euphrates, boosted by their tributaries, descended from the mountains to the nearly level Mesopotamian plain the rush of water slowed, allowing silts to be deposited on the riverbeds, raising their levels and causing periodic flooding. Sometimes these floodwaters swamped the works of man and buried them beneath yet more layers of silt. As a result of this silt deposition the Persian Gulf has retreated: the ancient town of Ur, excavated by Woolley, was at the head of the Gulf at the time of its First Dynasty, around 2700 BC, but is now over 200km inland. Taking his excavations down to almost 20m depth, Woolley found an ancient land surface of stiff green clay overlain by a metre of 'mud' which consisted largely of organic material. Remains of the earliest settlers in the area were found in this organic layer. Known as the al Ubaid period, and dating before about 3750 BC, the remains consisted of pottery in abundance, flints, clay figurines and flat (accidentally burnt) bricks. Immediately above this Woolley found a uniform stratum of 'clean silt', 3.3m thick, the origins of which he attributes to the middle reaches of the Euphrates River. This was overlain, in turn, by 5.5m of pottery fragments, with the kilns which produced them, embedded in the rising layer and, finally, a topmost layer, 6m thick, made up of eight layers of the remains of mud-brick houses. The Early Dynastic Period of Ur, dating from about 2700 BC, coincides with the top three housing layers.

Woolley identifies the natural silt layer as being the product of a 'Flood', from which may derive the story of the biblical flood. He also points out that the water needed to deposit this 3.3m layer would have been at least 8m deep, which, on the flat Mesopotamian plain, would have covered an area nearly 500km long and 160km wide, embracing the whole of the fertile land between the Elamite Mountains and the Syrian Desert. Only a few settlements, sitting on their own built-up mounds, could have survived. In order to deposit substantial depths of silts the inundation must have lasted many years, and may have been caused by rivers such as the Karun discharging massive amounts of silts into the Persian Gulf, forming a bar across it, behind which the silt-laden waters from the Tigris and Euphrates banked up and spread across the land.

It is apposite to ask why the ancient civilisations, which arose on the Mesopotamian plains, were so fixated with building enormously heavy structures on these soft alluvial soils. Strabo described the nature of the earth aptly: 'The soil is deep and soft; it yields so easily that the trenches and canals are choked or silted up and the plains near the coast form lakes and marshes filled with weeds.' Perhaps, even more pertinently, it might be asked how these peoples succeeded in building heavy structures on these soft soils.

Around thirty ziggurat sites have been identified in Mesopotamia, some of the sites having borne a succession of these massive structures, as the mud brick used in their construction deteriorated with time. Furthermore, they continued to be built over a period of more than three millennia, an early example having been found at Eridu, about 20km from Ur, and dating from the al Ubaid period, which was about 4000 BC, and the last, and most famous, the Tower of Babel built in Babylon around 600 BC.

The Tower of Babel derives its name from the biblical account of the establishment of a city and a tower by the families of the sons of Noah after the Flood. According to Genesis, Chapter 11:

> And it came to pass as they journeyed from the east that they found a plain in the land of Shinar; and they dwelt there. And they said one to another, Go to, let us make bricks, and burn them thoroughly. And they had brick for stone and slime had they for mortar. And they said, Go to, let us build us a city and a tower, whose top may reach up to heaven; . . . And the Lord came down to see the city and the tower, which the children of men builded . . . Therefore is the name of it [the city] called Babel; because the Lord did there confound the language of all the earth: and from thence did the Lord scatter them abroad upon the face of all the earth.

The tower referred to here is undoubtedly a ziggurat and its purpose clear enough – to enable man, or more likely his priests, living on the flat Mesopotamian plains to approach closer to God. The ziggurats were built mostly of sun-dried mud-brick with burnt brick facing, rather than entirely of burnt brick as implied here. The term 'slime' may well refer to bitumen.

Herodotus describes the ziggurat in Babylon as situated within the precincts of a temple area and states that it was still in existence in his time:

> it has a solid central tower, one furlong square, with a second erected on top of it and then a third, and so on up to eight. All eight towers can be climbed by a spiral way running round the outside, and about halfway up there are seats for those who make the ascent to rest on. On the summit of the topmost tower stands a great temple with a fine large couch in it, richly covered, and a golden table beside it. The shrine contains no image and no one spends the night there except (if we may believe the Chaldeans who are the priests of Bel) one Assyrian woman, all alone, whoever it may be that the god has chosen. The Chaldeans also say – although I do not believe them – that the god enters the temple in person and takes his rest upon the bed. There is a similar story told by the Egyptians at Thebes, where a woman always passes the night in the temple of the Theban Zeus and is forbidden, so they say, like the woman in the temple at Babylon, to have intercourse with men . . .

Koldewey's excavations confirmed the existence of a tower within the sacred temenos area of the inner city of Babylon dedicated to the god Etemenanki, but with base dimensions 90m square, rather than the 200m claimed by Herodotus. The importance of the priests of Etemenanki, and hence the tower, can be gauged from the fact that it was they who, on behalf of their god, bestowed the kingship of Babylon.

Architectural remains dating from about 3000 BC have been found by German excavators at Warka, near the southern border of Iraq, known in ancient times as Uruk and in the Bible as Erech. A mud-brick structure coated with lime and designated the White Temple, built on a pedestal 12m in height, may have been a forerunner to the ziggurat form, which, in effect, consisted of a succession of platforms of diminishing dimensions to give a stepped tapering form of tower, lending it stability while raising the incumbent temple closer to God. Architectural decoration at Warka included pressing into mud-plastered walls and columns small crayon-shaped pegs of baked clay, their blunt exposed ends coloured red or black, or retaining their natural yellow colour, giving elaborate mosaic patterns. Woolley found similar baked pegs in the pre-dynastic 'kiln stratum' of pottery fragments at Ur and concluded that they had been used to decorate a ziggurat of the Jamdat Nasr Period positioned on a high artificial platform surrounded by a steeply faced terrace wall.

The partially reconstructed ziggurat at Ur on show today dates from the time of the Third Dynasty king, Ur-Nammu, around 2700 BC. Buried inside it is an earlier structure built in the early dynastic period after the expulsion of the hated Jamdat Nasr usurpers from the site. Woolley believed this structure to be a ziggurat, but it is now thought more likely to have been a temple on a high raised platform, which again suggests a forerunner to the ziggurat form. Interestingly, Woolley found that the Jamdat Nasr builders used in their wall construction a flat brick, not unlike the modern brick, whereas their immediate successors in the early dynastic period used a much less practical plano-convex shape with a rounded top. This shape only lasted for a short period. Although enveloped by the ziggurat of Ur-Nammu, the dimensions of this earlier structure were not insignificant: its base measured 45m by 36m and, as with later ziggurats, it occupied part of a raised terrace enclosed by a heavily buttressed mud-brick wall 11m thick. The builders went to the trouble of importing limestone blocks to face the foundations of the wall to a height of 1¼m. The enclosed area within the walls contained other buildings of a religious nature dedicated, like the temple itself, to the Moon-god Nannar, the patron deity of Ur. A room near the base of the ziggurat appeared to be a kitchen, in which food may have been prepared and carried up to the temple to be offered to the god.

The first ziggurat of Ur, built in the time of the Third Dynasty king Ur-Nammu, occupied the site of the First Dynasty platform temple, which it consumed, and was positioned within a terraced area protected by a massive cellular mud-brick wall, built against the core of the old First Dynasty terrace wall. In the middle of the nineteenth century the British Consul in Basra, J.E. Taylor, struck by the size of one of the mounds when he visited Ur, initiated excavations into it. Although not penetrating to any great depth within it, he unearthed cylinders of baked clay supporting texts detailing the history of the structure. Dating from the time of Nabonidus, the last king of Babylon, the texts acknowledged Ur-Nammu and his son Shulgi as the original builders, but claimed that neither they nor any later kings finished it, this task falling to Nabonidus himself. In fact some 1,550 years separated Ur-Nammu from Nabonidus, who ruled in Babylon from 555 BC to 529 BC, and the ziggurat was not only completed by the Third Dynasty king but underwent various restorations over that long period of time.

Somewhat hampered by a government edict that he must not disturb any *in situ* brickwork, Woolley started his excavations at the site in 1922 and continued with the work for the next twelve winters. Sufficient of the structure remained intact for Woolley to be able to form a clear picture in his mind of its final form:

> In form the Ziggurat is a stepped pyramid having three stages. The whole thing is solid. The core is of mud brick (probably laid round and over the remains of the First Dynasty Ziggurat) and the face is a skin of burnt bricks set in bitumen mortar, about eight feet [2.4m] thick. The lowest stage, which alone is well preserved, measures at ground level a little more than 200 feet [61m] in length by 150 feet [46m] in width and is about 50 feet [15m] high; from this rose the upper stages, each smaller than the one below, leaving broad passages along the main sides and wider terraces at either end; on the topmost stage stood the little one-roomed shrine of the Moon-god, the most sacred building in Ur, for whose setting the whole of this vast substructure had been planned.
>
> On three sides the walls rose sheer to the level of the first terrace, but on the north-east face fronting the Nannar temple was the approach to the shrine. Three brick stairways, each of a hundred steps, led upwards, one projecting out at right angles from the building, two leaning against its wall, and all converging in a great gateway between the first and second terrace; from this gate flights of stairs ran straight up to the second terrace and to the door of the shrine, while lateral passages with descending flights of stairs gave access to the lower terraces at either end of the tower; the angles formed by the three main stairways were filled in with solid flat-topped buttress-towers.

Once Woolley had established the total form of the structure he realised that he had exposed the remains of a remarkable piece of architecture. The slope of the walls gave a

Diagrammatic reconstruction of the ziggurat at Ur.

much more pleasing aspect than would have been achieved by vertical walls, and contrasted in a pleasing manner with the flatter slopes of the triple stairs. All these slopes led the eye towards the all-important temple topping the ziggurat, while the horizontal planes of the terraces cutting across the inclined site lines removed any possibility of monotony in the structure. Puzzled by some discrepancies in his measurements, Woolley found on close examination that not one of the walls followed a strictly straight line, either from end to end or top to bottom. They had been given a slight convexity in both directions to impart a feeling of strength or, perhaps more correctly, to counter any illusion of weakness which may have emanated from a strictly planar wall fronting such a massive structure. Greek temple builders used the same optical trick more than 2,000 years later.

Rectangular openings spaced at regular intervals in the facing of the brickwork were found by Woolley to penetrate right through the burnt-brick casing deep into the mud-brick core, clearly indicating they had an engineering purpose, drainage, rather than being an additional architectural feature. He concluded that any water accumulating in the core would have caused swelling of the mud brick, quite possibly leading to bulging or even bursting of the outer walls. It is also possible that a build-up of water pressure under the incumbent weight of overlying structure could have caused weakening of the mud brick with resulting instability. But where did the water come from? As pointed out by Woolley it is unlikely to have been rainwater which caused the problem anticipated by the builders, because the terraces could have been paved with several layers of burnt bricks set in bitumen, which would have shed even the torrential deluges which occasionally occur in Mesopotamia. Woolley also dismisses the possibility that the builders were concerned about the substantial quantities of water in the mud mortar used in binding together the mud bricks of the core, as this would have dried out as construction proceeded. There is a possibility, however, that some moisture would have been retained in the mortar, and even in the bricks themselves, in which pressure would have built up under the increasing weight of overlying structure as the construction height increased, leading to possible instability. The builders may have known this from past experience.

Woolley took his cue for the *raison d'être* of the drainage ducts or 'weep-holes' from an inscription by Nabonidus, stating that, before undertaking the rebuilding of the ziggurat, he had to clear branches of trees from a temple complex situated at the base of the ziggurat at its south-east end. Woolley concluded that these could only have fallen from the ziggurat itself. He further concluded, therefore, that far from being paved with burnt brick, the terraces were covered with soil in which trees were planted. He conjured up an image of every terrace clothed with greenery, creating 'hanging gardens which brought more vividly to mind the original conception of the Ziggurat as the Mountain of God . . .'. This scenario envisages excess water from irrigation percolating down through the soil into the mud-brick core and eventually finding its way into the drainage ducts. But this poses some problems. In finding its way to the drainage ducts the water still had to pass through some mud brick, with the probability of softening it, and, furthermore, the mud brick might have suffered additional damage from the roots of the trees. Added to this the constant trickle of water issuing from the weep-holes would have caused unsightly permanent discoloration of the external wall facing.

A more satisfactory solution to establishing gardens on the terraces would have been to seal the top of the mud brick with layers of burnt brick in bitumen mortar, and then overlay this with first a drainage layer and then sufficient depth of soil to support the trees. At each end of the ziggurat Woolley found deep recesses in one of the buttresses extending from the edge of the first terrace to the ground, where it met an 'apron' consisting of a mass of brick waterproofed with bitumen with a slanted top to ensure a smooth splash-free run-off of water. It would have been a simple matter to direct the water reaching the drainage layer below the soil towards these recesses.

Towards the end of his reign Nebuchadnezzar, presumably having satisfied himself that he had the rebuilding of Babylon well under control, undertook a major rebuilding programme in Ur. He didn't finish it. In particular he appears to have done little or no work at all on the ziggurat, seemingly concentrating most of his efforts on the massive temenos wall. And there is no record of his three immediate successors, who ruled only briefly, continuing any of the work. This was left to Nabonidus, who came to the throne only six years after the death of Nebuchadnezzar. He found much of the lowest stage of the ziggurat intact (as it still is today) and needing only superficial restoration work. Everything above this was in a ruinous condition. Two and a half millennia later Woolley's excavation methods enabled him to recreate the plan and character of this portion of the first ziggurat.

Perhaps not surprisingly, Nabonidus placed a towering six stages on top of the existing base to create a ziggurat slightly smaller than, but otherwise almost identical to, the ziggurat in Babylon described by Herodotus. A temple for the god Nannar, faced with blue glazed bricks found by Woolley to be littering the site, topped the structure. Nabonidus retained the three first-stage staircases and the arched gate on the first terrace on which they converged, but above this the stages were accessed by zig-zagging staircases lying against the inclined walls. This arrangement differs a little from the spiral staircase described by Herodotus for the Babylon ziggurat.

Ziggurats were built over a much longer period than the other mass structures of the ancient world, the pyramids – some three millennia as opposed to barely one millennium. The ziggurats were an essential feature in the great city-states of Mesopotamia and there appear to have been no fewer than three in Assur, the ancient capital of Assyria situated on the Tigris River. No doubt rising civilisations felt a compulsion to built bigger and better structures than their neighbours or their forebears, whom they may well have vanquished.

Ziggurat construction not only continued over a much longer period of time than the pyramids in Egypt, but also extended over a wider geographical area, ranging from Tchoga Zanbil (near Susa, in ancient Elam, to the east of Mesopotamia) to Nineveh in northern Mesopotamia and Ur in the south. In the thirteenth century BC a powerful king, Untash-Napirisha, about whom little is known, ruled in Elam, a country which had close contacts with Mesopotamia. Inspired by the examples created by his neighbours to the west, and no doubt intent on exceeding them in sheer size, Untash-Napirisha built within his capital at Tchoga Zanbil a large ziggurat 105m square in plan and reaching through five levels to a height of some 50m. He had his name inscribed on many of the bricks to ensure that the Elamite god, Inshushinak, and future generations of priests and pilgrims, would have no doubt about who ordered its construction. The priests had access to the temple surmounting the structure in order to make suitable offerings and obeisances to the god, but pilgrims had to be content with making it to the first level. Unlike its Mesopotamian counterparts (with the exception of Nineveh), this ziggurat had rooms within its structure; but the entrances to these, featuring early examples of true arch construction, had been walled up.

Under the enlightened rule of Hammurabi (1792–1750 BC), Babylon grew from little more than a village on the banks of the Euphrates to one of the greatest cities of the ancient world. Hammurabi established, and presided over, a politically unified Mesopotamia. Under his successors, however, the unification was whittled away and in 1595 BC the First Dynasty of Babylon came to an abrupt end when the city was sacked by the Hittites, under King Murshilla I, who swept down from their northern Anatolian stronghold of Hattusha. Concerned at events back home, Murshilla returned quickly to his capital, but was too late to prevent a court uprising and his own assassination. He left behind in Babylon a power vacuum, which was filled by the Kassites, after a period of some uncertainty. Their first king, thought to be Agum II, assumed power in around 1570 BC. The Kassites had infiltrated into Mesopotamia from the mountains east of the Tigris, and into Babylon itself, during the First Dynasty of Babylon. There appears to have been no opposition to their assuming power; they had already adopted the local customs, language and religion. Agum is believed to have retrieved the statues of the god Marduk and his wife from the Hittites, and the restoration of their revered god would have endeared this king to the native Babylonians. So much so that the Kassite kings ruled Babylon for the next 400 years and established much of Mesopotamia as a unified political unit. They appear to have adopted liberal policies and presided over a settled land, but showed more interest in preserving existing legacies rather than in initiating major cultural or technical advances. They maintained close relationships with both Egypt and the Hittites through trade, exchange of gifts and even royal marriages.

They established a new capital at Dur-Kurigalzu, siting there, around 1350 BC, the Aqar Guf ziggurat, substantial remains of which can still be seen today just to the west of Baghdad. These impressive remains even today soar some 57m above the Mesopotamian plains, although the walls surrounding the lower portion are a modern restoration. Early travellers mistakenly thought it to be the biblical Tower of Babel. The remains seen today are characterised by horizontal ridges marking the locations of reed matting placed within the mud-brick structure to strengthen it, particularly against lateral instability. This technique, in principle, was revived only a few decades ago as 'reinforced earth' and is commonly used by civil engineers today as a means of stabilising and strengthening controlled earth fill, allowing it to be placed, with suitable facing, to a steep or even vertical slope. It was a technique known also to other builders in ancient times, having been used in some lengths of the Great Wall of China, where it consisted of sandy soil, the only locally available wall-building material.

In the Aqar Guf ziggurat reed matting, clearly visible in the denuded structure, was placed between every six layers or so of mud brick, and the structure further reinforced by cables formed from tough reeds running the full width of the structure from face to face, alternating in direction. In addition to enhancing lateral stability, the mats may also have

The partly reconstructed remains of the Aqar Guf ziggurat near Baghdad dates from the Kassite period in Babylon around 1350 BC. The ridges mark the layers of matting. Ducts were also provided to drain excess water from the structure.

been intended to smooth out uneven settlements, which would have caused unsightly distortions in the structure. Drainage ducts incorporated into the structure can also be seen.

In the period of Assyrian dominance of Mesopotamia, from about 880 BC to 612 BC, when Nineveh was sacked by an alliance of Babylonians and Medes, the Assyrian capital moved from Assur to Kalhu (modern Nimrud), then Dur-Sharrukin (now called Khorsabad after a local village) and finally Nineveh. In each of these they built towering ziggurats. Little remains of these, but excavations at Khorsabad have shown the dazzling visual impact these structures must have had, with the successive levels coloured white, black, rose, blue, vermilion, silver and gold. Carved veneers, coloured panels and statues adorned the walls. In his 1850 excavations of the Nimrud mound, Layard considered his most important find to be the 'strange pyramid [ziggurat] in the north-west corner of the site'. He estimated that the structure would have been more than 60m in height. Employing a workforce of thirty men he tunnelled into the base a distance of 25m, where he encountered a solid stone wall, which further tunnelling showed to have a thickness of 10m. He found within the base of the structure, unique in Mesopotamian ziggurats, a vaulted chamber, 30m long, 4m high and 2m wide. It was empty, but Layard found evidence that it had been broken into sometime in the past.

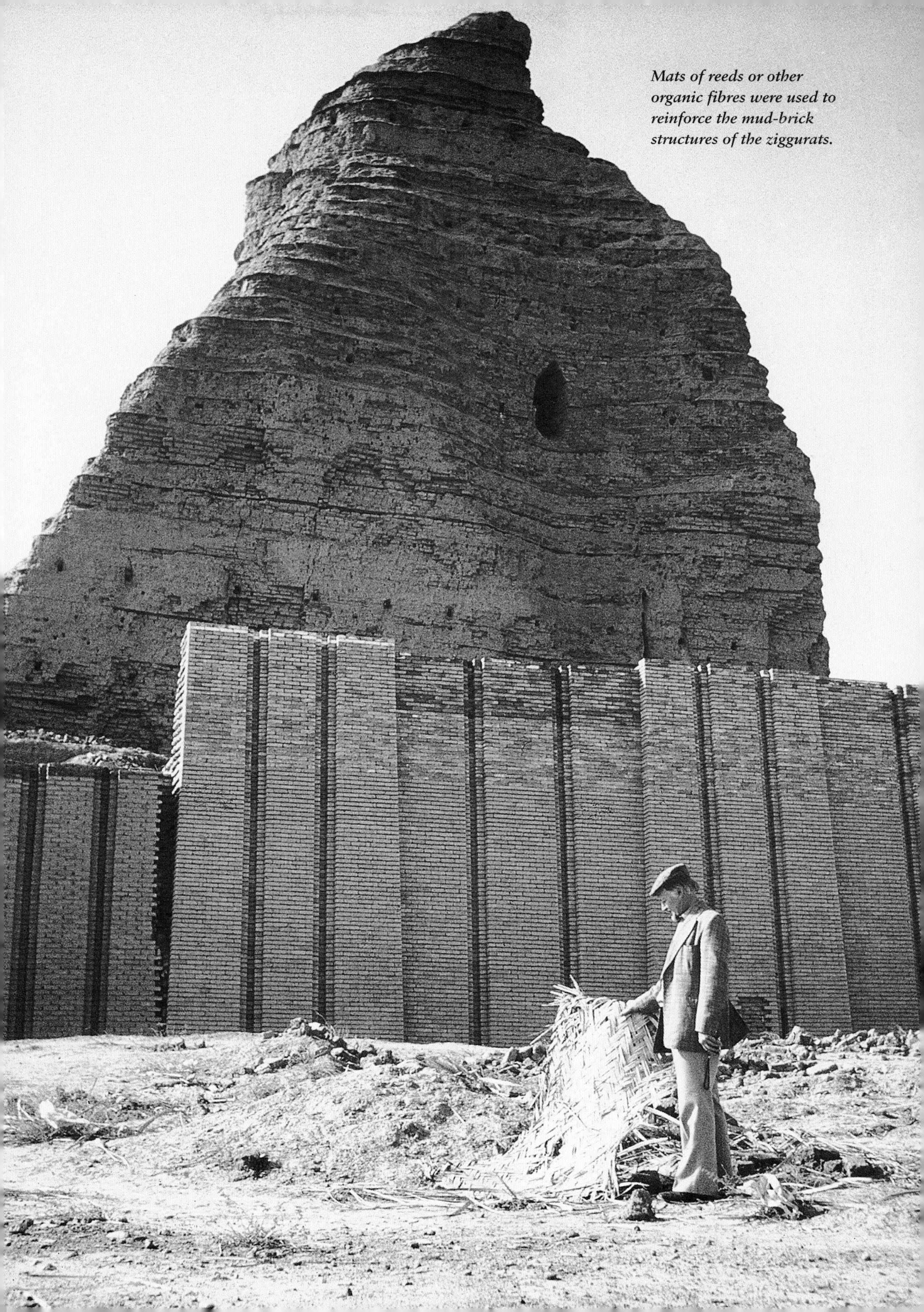

Mats of reeds or other organic fibres were used to reinforce the mud-brick structures of the ziggurats.

Probably built in the first century BC, the Pyramid of the Sun in Mexico rises to a height of 66m from a base 220m × 230m, the latter dimensions almost equalling those of the Great Pyramid at Giza. Although basically of earthen construction, an outer layer of stone, mortar and plaster gives it the clean linear lines of its appearance. Stone steps give access to the top, where a temple once stood.

Large earthen structures rivalling the Egyptian pyramids in size were built in the Valley of Mexico, near Mexico City, several hundred years before Christ. One of these, at Cuicuilco and probably constructed around 400 BC, was engulfed by a volcanic eruption about 2,000 years ago, but has been excavated and restored. It comprised a stepped circular structure with a base diameter of 145m, but flat topped with a height of only about 20m. It appears to have been abandoned before the eruption occurred.

In order to create a stable structure the builders at Cuicuilco first constructed an annular dyke made up of a mixture of clayey soil and stones, contained within an inner inclined wall of large stones and a series of similar outer-inclined stone walls. They then tipped readily available weaker clay into the circular hollow within the annulus. This represents an enlightened piece of geotechnical engineering, making best use of available materials. Finds of altar remains at its top, and also at an intermediate level, confirmed its usage for religious or ritual purposes and also indicated heightening of the mound at some time. Stone steps gave access to the top of the platform.

At the time Cuicuilco was overwhelmed, the great city of Teotihuacan, slightly to the north, had already been in existence for a hundred years or so, but its great period dated from AD 100 to AD 700, when it had a population approaching 200,000. A magnificent avenue, the Avenue of the Dead, linked the Pyramid of the Moon in the north of the city to the citadel and market place at its centre. The dominating feature of the city, however, was the Pyramid of the Sun, with a base 220m by 230m, almost equal to that of the Great Pyramid in Egypt and about half its height. It contains some 2.5 million tons of stone and earth, in a structure consisting of an outer stabilising skin, 15–20m thick, of adobe faced with an outer layer of stone, mortar and plaster. The bulk of the fill consists of various soils, suggesting the exploitation of several quarry sites, as well as stones and gravel. Figurines and potsherds have been found in some of the soils. Teotihuacan grew out of a simple agricultural settlement and there is no evidence that its inhabitants had the wheel or even sledges to transport the soil. They had no beasts of burden. In the oxygen-starved atmosphere at the elevation of Teotihuacan it was a prodigious task to build such a structure by human effort alone.

A platform or temple mound culture thrived in North America in the Mississippi River Valley and the South East between AD 700 and AD 1400. Some of these earthen platforms reached substantial proportions, an example in present-day Georgia having base dimensions 100m by 115m, a height of nearly 20m, side slopes averaging 32° and a total volume of 119,000 m³. Field evidence indicates that it was built in fifteen stages, each stage requiring some 6,700 man-days input. Even larger than this, the Cahokia Mound in present-day Illinois, also known as Monk Mound because Trappist monks once grew vegetables on it, is over 30m in height with base dimensions of 330m by 216m,

Emerald Mound near Natchez, built by forerunners of the Natchez Indians, dates from the period AD 1300–1600, late in the long period of mound building. Consisting in part of a reshaped hill, it is the second largest mound in North America, with a length of 235m, width of 133m and height of 11m.

Silbury Hill is 37m high and its base covers an area of 2.2ha.

considerably larger than the 230m square base dimensions of the Great Pyramid. Furthermore, the Cahokia Mound stood at the centre of more than a hundred smaller flat-topped mounds, with several hundred more within a few miles of it. It is possible, but not proven, that these mounds may have been the result of cultural exchange with, or even migration from, Mexico.

One of the great enigmas of the ancient world is the 37m-high Silbury Hill, the largest man-made mound in Europe and constructed at least 4,500 years ago. Silbury Hill has, not surprisingly, attracted much attention from antiquaries and archaeologists; but despite this, and the various tunnels and shafts that have been sunk into it, its purpose remains conjectural. The various excavations into Silbury Hill have shown conclusively that it was not built as a sepulchral mound and, despite its size, there has been no strong claim that it had astronomical usage. As pointed out by Burl, the location of Silbury Hill 'would have been a most inconvenient viewing-platform for astronomer-priests'. It is also doubtful if its primary purpose was linked to religious ceremonies or beliefs, although Dames has argued a linkage to the worship of a mother goddess. A plan view of the mound and ditches can be thought to represent a pregnant squatting woman, but this can only be perceived from the air and with not a little imagination, so it has not received strong acceptance by

Avebury area, showing superficial geology and the relative locations of Avebury, the source of its sarsen stones and Silbury Hill. Silbury Hill is situated on a spur of Middle Chalk projecting into the site from the south-west.

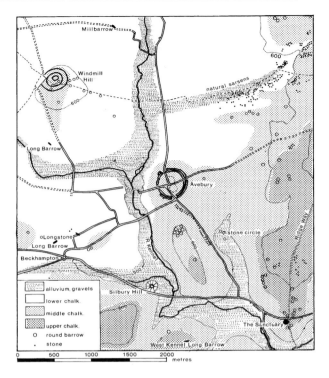

archaeologists. Worship of a mother goddess seems to have been stronger in other countries than in Britain. Perhaps more importantly, Silbury Hill is only one of a number of large mounds of Neolithic age in Britain not built to contain burials and with no astronomical significance. Important known examples are Marlborough Mound, the second-largest man-made mound in Britain, like Silbury Hill situated on the River Kennet and 5 miles downstream from it, and Hatfield Mound, now demolished, but previously occupying a site midway between Avebury and Stonehenge. A particularly significant example may be Gib Hill, situated close to Arbor Low, a henged stone circle in Derbyshire reminiscent of Avebury, although much smaller. Arbor Low has its own smaller relative in nearby Bull Ring, which also has an adjacent non-sepulchral mound. In Brittany the large tumulus, St Michel, near the Great Stone Alignments of Carnac, has some minor burials within it, but these are insignificant in proportion to the mound itself and evidently not the primary motivation for its construction.

If Silbury Hill is not a sepulchral barrow, and served no religious or astronomical purpose, why was it built? It is inconceivable that such enormous effort would have been put into building it unless it were to perform a specific and highly important function. Is there any evidence that might reveal its real purpose? While there may exist no direct evidence, the solution needs to satisfy a number of considerations.

1. In view of its close proximity to the Avebury stone circle, Silbury Hill must be closely associated with it, as stressed by Hoare in the nineteenth century and many others since, and it may be no coincidence that this largest mound is adjacent to the largest stone circle in Britain. Both date from the Neolithic period and some organic material from the base of Silbury Hill has indicated a corrected Carbon-14 dating of about 2700 BC. However, this is not necessarily the date it was placed at this site and the construction may have continued over a very long period. The Avebury circle is even more difficult to date precisely, as is the sequence of operations in its construction. It is possible that the two smaller circles within the great circle were constructed first to allow some usage of the site from an early date.

Viewed from the top of the Avebury embankment, Silbury Hill is partly obscured by the flank of Waden Hill.

2. Unlike barrows raised for sepulchral purposes, which were usually sited on or close to hilltops, Silbury Hill is situated in a valley at the foot of a chalk hill. The nearby Kennet Long Barrow, for example, occupies a very prominent ridgeline position. While the location of Silbury Hill may have been for geological reasons as discussed below, it would surely have been placed in a much more prominent position had it been built for religious or astronomical purposes. There seems to have been a deliberate attempt to reduce its visual impact as much as possible, particularly from Avebury itself, from most parts of which the mound cannot be seen, partly because of Avebury's earthworks and partly because of Waden Hill, which separates the two sites. Malone makes the point that 'it would not have been visible from within the Avebury circles when the banks were at full height'.

3. Excavations (now silted up) around the base of the mound provided chalk and soil for its construction, but, as observed by Malone, these trenches 'cannot have provided all the chalk needed to build the hill and other areas around must have been exploited'. The total volume is about 0.33 million m³, of which the excavations account for about 0.17 million m³. However, the lower few metres of the hill are made up partly of a natural chalk spur and, as a rough estimate, about 0.12 million m³ may have been imported. Where did this come from?

Despite its size and relatively steep slope of 30°, Silbury Hill has retained its stability for some 4,500 years because it is a carefully engineered structure, designed by people who learnt from experience the problems associated with the stability of earthen materials. The various diggings into Silbury Hill have failed to find a grave (simply confirming Hoare's prediction made in 1812), but they have revealed some details of the inner structure of the mound that explain its stability and perhaps its true purpose. From an engineer's point of view this gives a more revealing insight than any burial into the lives and technological know-how of the people who built it, particularly with respect to their exploitation and protection of the land.

Three principal phases have been identified in the construction of the mound, deduced primarily from the tunnel excavations directed by Professor Richard Atkinson in 1968 for a BBC documentary. Details of Atkinson's findings were published in 1997 in Oxbow Monograph 74, *Sacred Mounds, Holy Rings* compiled by Alasdair Whittle and other contributors, where the point is also made that although these phases encompass the main features of the mound (in so far as it has been possible to observe them), an open mind should be kept on the actual sequences of construction or, indeed, on the period of time over which the construction may have extended.

View in the direction of Silbury Hill from within henge area. Silbury Hill is totally obscured by the henge embankment.

The first phase consisted of placing on a spur, projecting from the adjacent hillside, a layer of gravel, followed by a stack of *turves* topped by four consecutive layers of soil, clay, chalk and gravel, resulting in a drum-like mound, 4.5m high, within a circular staked-out area with a diameter of about 37m.

The second phase consisted of placing primarily chalk rubble to form a mound of roughly conical shape with flat slopes, having a base diameter of about 73m and reaching a height of perhaps 17m, and covering the first-phase mound. The chalk rubble was placed in flat sloping layers, and surface layers of harder chalk may have protected the outer slopes.

The final phase saw the mound taken to its final height. Trench excavations on and near the top of the mound have indicated that it is made up of a cellular structure consisting of a series of circular concentric and radial walls of harder chalk blocks or rubble with softer chalk rubble infilling. If, as is likely, this structure extends for the full height, or a substantial part of the height, of the mound, it is a remarkable piece of civil engineering and very successful, too, as the mound has remained stable with slopes of 30° or slightly greater, which the softer chalk, without the retaining walls, could not have sustained.

A fourth phase may be added to the above: the placing of topsoil and *turves* on the outer surface of the mound to give it a smooth exterior, and quite possibly to reduce,

Excavations in and near the top of Silbury Hill have revealed concentric and radial chalk walling, forming cells retaining softer chalky rubble. The construction materials were raised to the progressively higher working levels up a spiral ramp, which was subsequently covered with top soil or turves to give a smooth exterior and reduce the visual impact of the mound.

as much as possible, the visible impact of the glaringly white chalk making up the structure. In fact some small undulations can be seen in the external profile of the mound, once thought to indicate that the mound was built up with a series of external steps or terraces about 3–5m wide. It is now recognised that these are not terraces, but a spiral ramp, which is entirely logical as it would have given much readier access for the workmen to haul the rubble to working level for placement than scaling ladders up steep-sided terrace slopes as suggested by some leading writers. Although the Oxbow Monograph refers to the sub-surface feature as terraces, it notes that the first terrace, near the top, is 'not level around the circumference of the mound'. The recognition of the true origins of the surface undulations has, unsurprisingly, led to suggestions that the priests, or their like, used the spiral ramp to make ceremonial processions to the top of the mound. There is no evidence for this, and only highly doubtful evidence at best to support claims that the mound had any ritual significance at all. If the ramp served this purpose, why was it filled over? As suggested above, the ramp may well have been filled over with *turves* to reduce its visual impact, but this would also have ensured that the mound would more readily shed rainwater and thus contribute to its stability.

The proximity of the largest ancient artificial mound in Britain (indeed in Europe) to the largest stone circle is unlikely to be a coincidence, and strongly suggests a close linkage between the two. At first sight, however, this seems in conflict with the placing of the mound in a valley location so that it could not be seen from within the Avebury circle, and the suggestion that covering the mound with *turves* was, in part at least, to lessen its visual impact as much as possible. Adding these observations to the fact that quite extensive exploration has produced no evidence of any burials, that its location is completely wrong for any use as an observatory, and that there is, at best, only scant and unconvincing evidence of its creation as a ritual monument leaves only one obvious possibility – that Silbury Hill was an unwanted by-product from the construction of the Avebury circles.

Avebury henge, inside its surrounding ditch, is almost circular, with a mean diameter of 348m. Originally 7–10m deep with top and bottom widths, respectively, of about 23m and 4m, but now half silted up, the ditch provided the chalk for an embankment 17m high lying outside it, now reduced by erosion to a height of 4–5.5m. These are massive earthworks in themselves, although dwarfed by Silbury Hill. Chalk walling, similar to that in the Silbury mound, and using harder chalk from the base of the ditch, was used in places to stabilise the inner face of the bank. There are four causewayed entrances passing through the embankment and across the ditch into the site.

Some 247 stones stood within the Avebury henge, 98 of which, generally 3–4m in height and up to 40 tons or more in weight, were erected just inside the ditch, comprising the outer circle, and the remainder in two inner circles, both, in themselves, having greater diameters than the outer circle of Stonehenge. The northern inner circle originally consisted of 30 stones making up the outer perimeter, with a diameter of almost 100m, and contained within it a structure known as the Cove, consisting of three huge stones arranged on three sides of a square. The southern, slightly larger, circle had a perimeter of 32 stones and at its centre a huge stone 6.4m high, known as the Obelisk, with an arrangement of smaller stones attending it. Very few stones of the two inner circles survive today and less than a third of the outer circle stones. Two stone-lined avenues led away from the henge: Beckhampton Avenue to the west and West Kennet Avenue, some stones of which survive, to the south-east. It has been estimated that these two avenues may have numbered over 300 stones between them.

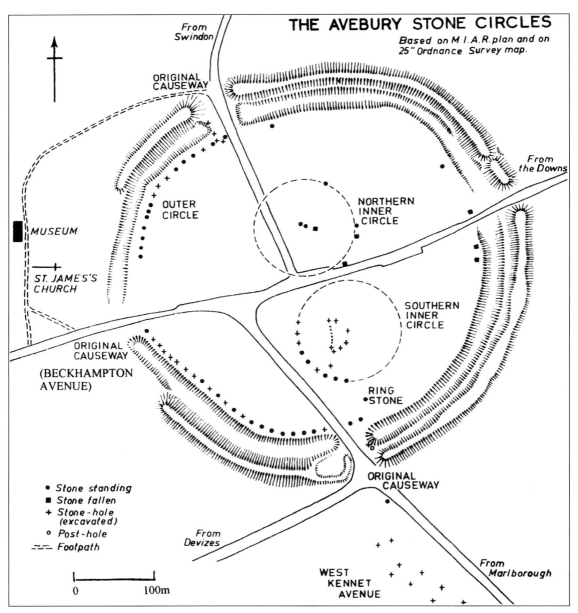

Plan of Avebury site.

The stones at Avebury derived from nearby Marlborough Downs to the north-east, as did the Stonehenge sarsens. Exposed stones, or 'wethers', can still be seen there. There appears to have been some attempt to select stones by their shape, as the surviving examples tend to be in the form of tall pillars with vertical sides or broad diamond or lozenge in configuration, standing on one corner. Unlike the Stonehenge sarsens, however, no attempt was made to dress the stones, making them more awkward to handle and transport than those at Stonehenge, but with the compensation that they only had to be transported for much shorter distances, perhaps up to 2km or so.

It would not have been possible to attach timber pieces to these rough, undressed stones, to form a cylinder and roll them, which might have been possible with the Stonehenge sarsens. In order to haul them to the Avebury site they would have to have been mounted onto a timber sledge. This would have been a formidable task in itself, but not impossible with levers and ropes. Frozen ground would have been ideal over which to haul the stones, but the climate at that time may have been slightly warmer than at present, so this option would not have been generally available. In order to reduce friction a sledge would have required a specially prepared and compacted haulage track with

Avebury stones. Lifting the Avebury stones, some weighing up to 40 tonnes, to the vertical position would probably have been accomplished by tipping them over the edge of a ramp or embankment. Various proposals have been presented employing this technique (e.g. Malone, Parry, Richards and Whitby, Simpson). Silbury Hill may consist largely of this chalky ramp or embankment material and causeway material removed at the end of construction.

greased cross-pieces of timber, like railway sleepers, set into it. A more likely alternative would have been to pull the sledge over rollers, but this is only feasible if the rollers are uniform in diameter and if the haul track is very firm, very even, and free of any other than very gentle slopes. Either case of sledge usage would have required a specially prepared haul track using carefully selected local chalk capable of being compacted to give a firm, even surface, set on an embankment averaging perhaps 2–3m in height to ensure gentle slopes from source to site, and which would quickly shed water to prevent softening of the surface. The fall in land elevation from the source of the stones to the Avebury site was around 50m, so the haulage track could have been built with a uniform slope of about 1 in 40, which would have been manageable, whether drawing the sledge over cross-pieces or rollers, and indeed would have eased the task of the hauliers.

Raising the stones into the vertical position once they reached the site presented no less a task than their transportation to the site. Maximum advantage can be taken of a stone's own weight by carefully positioning it and tipping it over the front of a ramp or the edge of an embankment; this method was shown to be feasible in full-scale Stonehenge field tests by Richards and Whitby, in which the test block to be tilted was balanced on the front edge of a strengthened ramp and tilted over the edge to an angle of 70° by sliding forward a weight resting on the block. It was then pulled to the vertical position using ropes and an A-frame. An alternative method using only ropes and a lifting frame positioned on the ramp behind the stone, and tested at model scale, has been proposed by the writer. Other methods using ramps, together with levers and ropes, have been proposed (e.g. by Malone).

In view of the number of stones within the Avebury henge, comprising the outer and two inner circles, and the others within the inner circles, it would have been simplest to cover the whole site, or most of it, to a depth of perhaps about 2m with well-compacted chalk rather than having either a multitude of individual ramps or having constantly to reposition ramps. This would have given maximum manoeuvrability on site and allowed stockpiling of stones. Silbury Hill itself is testimony to the ability of these Neolithic workmen to quarry, transport and place these large amounts of earth and rock fill. And it may well be that the silted-up quarry often described as 'encircling' Silbury Hill provided some or all of the earth and chalk used in the construction of Avebury, and it was only after completion of this work, which may have taken centuries rather than decades, that this fill material, now a waste product, was taken back to its source to create Silbury Hill. Disposing of this essentially sterile material would have posed a major problem: disposing of it on adjacent hard-won farmland was not an option; placing it back in the quarry whence it came might have been possible but for the fact that it would have been partly, or wholly, silted up – as it is now – with hillwash and stream deposits. It might also have been a very soft, waterlogged area, so placing the unwanted material over this silted-up quarry may not have been an option, and the workmen chose instead to place it, for the most part at least, on the harder rounded spur of rock projecting into the site from the south-west, where the rejected superficial material from the quarry excavation had already been placed much earlier. It is possible that this primary mound also contains some earlier superficial material cleared off the Avebury site. As suggested below, some of the mound may in fact be sited over part of the quarry.

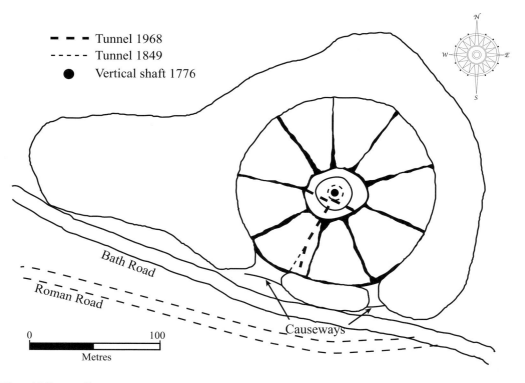

Plan of Silbury Hill.

The description of the quarry 'encircling' Silbury Hill implies that it stops in front of, and does not extend below, the mound; but if the purpose of the quarry was to provide fill for construction of the Avebury circles it might be expected that the quarry in fact encircled the rounded spur on which the mound is situated. It could be that this particular quarry area was chosen because it could not be seen from the Avebury henge, that it was not considered good farming land and because it yielded a chalk fill of a texture and hardness easily quarried and readily worked into a compact mass suitable for ramps and causeway embankments. The chalk of the spur itself would have been generally too hard for these purposes, but would have been eminently suitable for special uses, such as for chalk walling in the mound and possibly embankment edges at locations where the Avebury stones were tipped up into their vertical positions. The chalk walling in Silbury Hill played the role of containing the softer fill brought back from the Avebury construction works and stabilising the mound so that the softer fill would not wash away to spoil large areas of farmland and contaminate local streams. It has been highly successful for some 4,500 years and constitutes a remarkable piece of environmental engineering. It is truly an environmental icon.

If the quarry encircles the spur, rather than the mound, then part of the quarry may lie under the mound; unfortunately the many explorations of Silbury Hill have provided little useful information on this. The tunnel driven by Atkinson in 1968 has provided most of the known information at or near the base of the mound, but was driven from above the spur and not from the main quarry area. Nevertheless, observations taken in the tunnel indicated that some harder chalk was taken from the spur and, as shown in the Oxbow Monograph, the excavations do lie partly under the mound. In 1886 pits put down by Pass in the main quarry area revealed a filling of chalk silt averaging 4.6m in depth, but increasing in depth to 6.4m at the base of the mound, which suggests that the quarry may extend for some considerable distance underneath the mound.

If Silbury Hill is composed of waste fill from the Avebury construction site, it should be possible to reconcile the volume of the mound with the amount of fill that may have been used in transporting and erecting the stones. The volume of fill in the mound is about 300,000 m³. If a 2m depth of fill was placed over a substantial area of the Avebury site, this could have amounted to 150,000 m³. The embanked causeway along which the stones were transported, assuming it to be 2km long, 15m wide and averaging 2m in depth, accounts for another 60,000 m³. The remaining volume is easily accounted for: by earthworks used to erect the stones for the Beckhampton and West Kennet Avenues, by the primary mound in Silbury Hill and by some, at least, of the chalk walling quarried specifically to be used in the mound itself.

It is probable that some of the fill used in the construction of the Avebury circles and subsequently placed in Silbury Hill did not come from the Silbury quarry. One possibility was the Windmill Hill causewayed camp situated 2km north-west of Avebury and pre-dating Avebury and Silbury. A system of concentric ditches, the outer up to 2.5m deep and encompassing an area of 8 hectares, surrounded the central area. The excavated chalk forming banks along the edges of the ditches has disappeared. This could be due to erosion, but if the banks still existed at the time of the Avebury construction, they would have been a tempting source of fill for erecting the Avebury stones. One of the entrances to the Avebury henge is oriented towards Windmill Hill, perhaps to give ready access to the works. In his excavations in the top of Silbury Hill Atkinson found a potsherd of flint-gritted Windmill Hill ware, which gives some additional credence to the possibility that some of the fill came from this source.

Radiocarbon dating, discussed in the Oxbow Monograph, provides disappointingly little evidence in trying to unravel the story of Silbury Hill, but points to the placing of the *turves* in the primary mound having taken place around 3000–2800 BC. This is in contrast to the dating of around 2300 BC or later for a deer antler found in the base of a quarry excavation outside the southern edge of the mound. The quarry excavation may have provided harder chalk for the retaining walls in the mound. As the placing of the *turves* probably marked the start of the main quarrying operation, these dates are not contradictory with the construction of the mound taking place some centuries later, after completion of the work on the Avebury site.

The vast number of Neolithic stone circles and other standing stones in the British Isles points to the likely existence of a group of experts, akin to the Masonic guilds of the Middle Ages, hiring out their closely guarded expertise. Although the precise construction of these monuments would have been dictated by local conditions such as the nature and availability of megalithic stones and earth and rock fill, and the availability of timber, it should be possible to detect some similar characteristics. Of particular interest, in relation

View of Arbor Low showing the fallen stones.

to Avebury and Silbury Hill, is Arbor Low in the Peak District, about 200km north of Avebury in a direct line. A number of writers have noted the similarities between the two sites; particular attention is drawn to Gib Hill, an earthen mound situated 300m distant from the Arbor Low henge and clearly associated with it.

The Arbor Low henge, inside its ditch and external bank up to 2m high, is slightly oval, with dimensions of 83m by 75m. All the limestone blocks, up to 8 tons in weight, making up the circle, or more correctly egg-shaped ring 37m by 42m, are now lying in a regular manner on the ground, but almost certainly once stood upright, as attested to by seven stumps of broken uprights. As with the northern circle in the Avebury henge, a cove of three larger stones occupied the centre of the Arbor Low ring. Gib Hill is a Neolithic mound 35m long, 20m wide and 1m high, primarily formed of local red clay, containing, in itself, no burial, although it is topped by a later Bronze Age barrow with a small superficial limestone cist in which cremated remains were found. The earth fill making up the Neolithic mound, and which may well have been initially excavated as ramp material

Gib Hill. A raised track linking Arbor Low to Gib Hill took a curved path, possibly around a stand of trees left in place or specially planted so that the mound could not be seen from the stone circle.

for erecting the Arbor Low stones, almost certainly came from quarry pits lying 50m or so to the side of Gib Hill, the indentation of which can still be seen.

The passage of the earth fill between the henge and Gib Hill seems to have been along a slightly embanked trackway or 'avenue', parts of which can still be discerned. Interestingly, however, it heads away from the henge in a south-south-westerly direction, which would miss Gib Hill by 100m, but, on coming abreast of the mound, it turns westward towards it. The reason for this may have been to ensure that there was no visual contact between the henge and the mound by leaving a stand of forest trees between them, suggesting that Gib Hill, like Silbury Hill, was simply a mound of waste material.

As observed by Pitts, many of the henge sites in Britain have associated with them non-sepulchral mounds or barrows. One example, lying slightly to the north of Arbor Low, is the Bull Ring, from which the standing stones have long since been removed, but a substantial mound still remains about 30m outside the henge bank. Although the stones are severely weathered, the Rollright Stones near Oxford is one of the best-preserved circles in the country. About 73m north-east of the circle is a large stone known as the King Stone and beside it are the remains of a Neolithic mound. Several other examples can be cited, but such mounds are absent from the sites of some circles, because they have been ploughed into the ground. An example of such destruction is the obliteration of Hatfield Mound, situated in an enclosure near the village of Marden, some 12km south of Avebury. It is thought to have been about 150m in diameter and 7m high. Further afield, the large tumulus of St Michel at Carnac in Brittany, near the Great Stone Alignments, contains some minor burials, but these are insignificant in comparison to the size of the mound and were clearly not the primary reason for its creation.

Neolithic communities were building huge timber structures within henges both before, and contemporary with, stone circles such as Avebury and Stonehenge. Examples include Woodhenge and Durrington Walls, close to the Avon River and near Stonehenge, and The Sanctuary near Silbury. Woodhenge and The Sanctuary consisted of six concentric rings of massive timber posts, almost certainly supporting some form of roof structure, the outer circle at the former spanning a diameter of 38m, while Durrington Walls comprised two structures termed the Southern and Northern Circles. The dimensions of the postholes marking the positions of the vertical posts attests to the enormous size of the timbers used. For example, the largest postholes at Durrington Walls were found to be no less than 2.2m in diameter and 2.9m deep, indicating posts perhaps up to 8m in height and weighing as much as 12–15 tons. The task of raising these timbers to their vertical positions would have been comparable with raising huge megaliths and would probably have been accomplished using earthen ramps or embankments as proposed above for the stone circles. It is conceivable that the fill used at The Sanctuary eventually became incorporated into Silbury Hill. It might be significant that the demolished Hatfield Mound lay close to the sites of Woodhenge and Durrington Walls. Another of these large timber structures at Mount Pleasant in Dorset has a large mound, known as Conquer Barrow, situated on the western bank of its enclosure. In this case, if it does contain the waste material from the construction of the building, no attempt was made to hide it from view.

The amount of fill material needed for the construction of Stonehenge would have been very much less than that required for Avebury, but, nevertheless, the waste material would have formed a substantial mound. In view of the fact that Stonehenge was almost certainly the most sacred site in the country, the builders are likely to have been anxious to remove the waste as far away as possible. The resulting mound may long since have been ploughed into the ground. One possibility is Hatfield Mound, mentioned above with respect to Woodhenge and Durrington Walls. It was only a few kilometres from Stonehenge and accessible along the navigable Avon River. Another possibility is the 20m-high stepped chalk mound in the grounds of Marlborough College, some 28km to the north, the purpose of which has not yet been established. Transporting the waste this distance would hardly have been a problem to people who had brought bluestones to the Stonehenge site from Wales and huge sarsens from Marlborough Downs, 40km to the north.

7

BUILDINGS FOR THE GODS

Religion, or spiritual belief, has always been a strong motivating force throughout mankind's existence, and some of the greatest buildings in the ancient world, as today, were built for religious purposes.

The two dozen or so temples in Malta, incorporating huge megalithic blocks of limestone and dating from 3500 BC to 2500 BC, are among the earliest religious structures of stone ever built and are unique in design. They are contemporary with, or pre-date, the megalithic structures of Western Europe such as those at Avebury, Stonehenge and Carnac. The farming-based community that built these temples, thought to have arrived in Malta around 4000 BC, found they had an abundance of workable limestone available for their construction and developed the techniques to quarry this rock, and to transport and raise huge megaliths. The basic unit of temple design, differing from anything else in the ancient world, had a perimeter wall, up to 6m high, of massive upright megaliths or orthostats, usually placed alternately with faces and edges facing outwards, forming a rough 'D' shape in plan; but having a slightly concave façade with all faces exposed, and an entrance through it giving access to a number of chambers, either sub-elliptical or trefoil in plan. Chamber walls consisted of roughly squared uprights surmounted by large horizontal blocks. Some of the chambers had short interconnecting passages, in some cases formed by enormous uprights. More elaborate temples were built adopting or merging multiples of the basic 'D' form: that at Mnaidra, for example, having a double 'D' form, with two entrances.

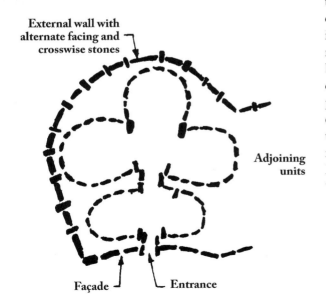

External wall with alternate facing and crosswise stones

Adjoining units

Façade   Entrance

A common form basic unit making up the Neolithic temples in Malta, having concave façade and trefoil and sub-elliptical internal walled chambers. The temples comprised two or more such basic units.

Frontal view of a small chamber in the Mnaidra temple in Malta, showing the rectangular entrance cut through a single stone slab. Orthostats with horizontal capping stones forming the adjoining wall can be seen on the left. The stones exhibit decorative indentations. (From Hammerton)

The basic geological structure of the island, of Miocene Age, consists of a lower layer of coralline limestone, with a layer of globigerina limestone resting on it and separated from an upper layer of coralline limestone by layers of a blue clay and sand. The coralline limestone, much the harder of the two rocks, could nevertheless be broken away in slabs, which, by exploiting natural fissures, could be reduced to manageable sizes and used, unhewn, mainly for external walls which were exposed to weathering. Some internal walls were formed of piled-up rough coralline blocks. The softer globigerina limestone, on the other hand, could be readily quarried and worked using stone tools and antler picks, making it suitable for internal walls and other features, the walls consisting of well-cut slabs set upright as orthostats, supporting upper courses of stones set horizontally. Internal walls were rendered with a lime plaster painted red. All walls had double faces – inner and outer – and the gap filled with earth or rubble, or occasionally left unfilled to provide subsidiary chambers. Doorways usually consisted of parallel orthostats topped by a lintel, but in some cases were formed by cutting a rectangular hole in a single upended slab. Large exposed blocks of globigerina limestone were frequently elaborately decorated with patterns of surface indentations or surface carving, often with spiral motifs, reminiscent of other Neolithic monuments, notably Newgrange in Ireland.

The floors consisted of a crushed limestone concrete. In some of the temples the upper walls of the chambers are corbelled inwards, probably to reduce the distance timber beams had to span in supporting a flat roof, waterproofed with layers of brushwood and clay.

While perhaps not without some religious significance, most stone circles in Britain may have been primarily secular meeting places, where, for example, local tribal leaders would settle their differences on matters such as land usage. They may also have served as market places for the exchange of produce and animals. Stonehenge, on the other hand, certainly had overwhelming religious significance and it is likely that only the priests had access to it. It was a temple. The priests would have handed down, to their chosen successors through the generations, the knowledge of how to exploit the structure to predict cosmic events and so enable them to maintain a powerful influence over the lives of the people: the people, in return, would have received both spiritual comfort and advice, or perhaps even instruction, on practical matters such as the right times to plant their crops.

The term 'henge' usually relates to a parcel of land – about 110m in diameter in the case of Stonehenge – enclosed, in ancient times, within an encircling ditch and an embankment,

Spiral decorations were common in Malta temples, as elsewhere in Neolithic structures, notably Newgrange in Ireland. (From Hammerton)

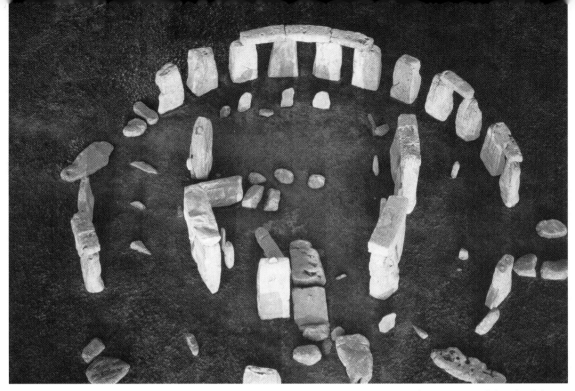

Stonehenge differs in a number of respects from other stone circles in Britain, including the capping with lintels of the outer stone circle, the massive trilithons, the use of stone from two entirely different sources and the degree to which the stones were dressed. Its clear orientation towards the summer and winter solstices strongly points to a ritualistic function, while other stone circles may have had more secular functions.

The smaller bluestones at Stonehenge, up to 4 tonnes in weight, derive from the Preseli Mountains in Wales, 250km away, while the larger sandstone sarsens and lintels, up to 40 tonnes in weight, came from Marlborough Downs, 40km to the north.

formed from the soil and rock excavated from the ditch. The site itself may have had sacred connotations and perhaps even served as an observatory long before the introduction of the first stones around 2500 BC, as radiocarbon dating has shown timber posts to have been erected here as early as about 8000 BC. The ditching and embanking to form the circular enclosure site took place around 3000 BC, with timber structures occupying the site until the introduction of the first stones, the bluestones, around 2500 BC. Also associated with this early period, or Phase 1, of Stonehenge are the heelstone – a standing stone 6m high and weighing some 35 tons – placed on the line of the entrance about 30m outside the henge, and a series of fifty-six holes, called 'Aubrey Holes' after their discoverer, just inside the inner bank. The purpose of the holes is uncertain.

The stones of Stonehenge differ from those forming other stone circles in originating from two entirely different outcrops, as well as in the long distances of the outcrops from the site, and also in the degree to which the stones were dressed to a regular shape, particularly the larger stones – the sarsens – and the lintels. It is the only circle with lintels topping the stones, but their purpose is unclear as there is no evidence that they supported a roof structure over the stones.

Credence for the ritualistic or spiritual significance of Stonehenge derives from the orientation of its main axis, encompassing the avenue accessing the site, the heelstone within the avenue and the gap between the largest trilithon stones forming part of the horseshoe of sarsens within the outer sarsen ring. The orientation of this axis is towards the summer solstice – the point of sunrise on the horizon in mid-summer – but it is equally towards the winter solstice – the point on the horizon of mid-winter sunset. Although the former is generally assumed to be the more important, and celebrated by certain sects even today, there is an equally strong argument for the importance of the winter solstice, marking the time when the land starts to come to life again. Four sarsens, known as the Station Stones, were erected outside the main stone circle, but just inside the embankment; the lines joining them form a rectangle with the long sides oriented north-west and south-east, and the short sides north-east and south-west. These orientations coincide reasonably well with the outer limits of the 18.6-year moon cycle, strongly suggesting the use of Stonehenge as both a solar and lunar observatory.

Around 2550 BC some eighty-five stones formed of igneous rock (i.e. deriving directly from molten lava), known as bluestones, and mostly weighing 3–4 tonnes, were erected on the site in an uncertain configuration, but probably forming a double concentric, partly circular, arrangement. Although showing some variation in stone type they all derived from an outcrop in the Preseli Mountains in south-west Wales, 250km from the Stonehenge site. There is a body of opinion that states that glaciers conveyed the stone to the vicinity of the site, but this raises the question of why no other scatterings of the stones have been found. It is more likely that they arrived at the site through human effort, the most common suggestion being that they were hauled on sledges over land to Milford Haven and then shipped up the Bristol Channel to access the Rivers Avon and Wylye, along which, together with some more overland haulage, they eventually reached the site. This would have presented immense problems, as evidenced by the failure of a millennium project attempting to duplicate this in 2000. The stone sank in the Bristol Channel. A more straightforward method of transportation would have been to lash shaped pieces of timber to the stones to

Field tests by the author have shown that a 3-tonne concrete block, fitted with shaped timbers to give it a cylindrical shape, could easily be rolled up a 1 in 7 slope, suggesting that the bluestones could have been transported from Wales entirely overland in this way. Although this would have been much more challenging, it is possible that the same technique could have been used to transport the sarsens.

give them a rough cylindrical shape, which would have presented no great difficulty in view of the slabby nature of the stones and the unquestioned skill of the Neolithic woodworkers. This would have allowed the stones to be moved entirely over land by rolling them. The writer has demonstrated the ease with which 22 men could roll a 3-ton concrete slab up a 1 in 7 slope close to the Preseli outcrop, proving an overland route to have been entirely feasible. Rivers could easily have been crossed at ford points or shallow sections, firming up the bottom with stones if necessary: buoyancy would have balanced much of the weight of the stone and its attached timber pieces and so facilitated the crossing.

The erected bluestones remained in place for less than a century before being dismantled and stockpiled, and huge sarsens of hard indurated sandstone, weighing up to 40 tonnes, brought to the site from Marlborough Downs, nearly 40km to the north. Despite the lesser distance, transportation of these stones to the site would have been much more difficult than for the much lighter bluestones. Richards and Whitby demonstrated in 1997 the possibility of moving a 40-tonne concrete block by sliding it on timber rails, but limited their efforts to hauling it up a 1 in 20 slope, leaving unanswered the problems of crossing the River Kennet and negotiating the steep Redhorn Hill. All of these difficulties would have been greatly reduced or eliminated if the rolling technique had been used. This would have reduced, by 50 per cent or more, the required labour force of up to 200 for the 1 in 20 slope. Either method could conceivably have used animal power, rather than human power, in the form of oxen or cattle as draught animals, but again the smaller number of animals required to roll the stone would have made controlling them easier.

Thirty dressed sandstone sarsens, weighing up to 25 tonnes each, were erected in a 30m-diameter circle, each to a height of about 5m with 1.2m buried in the ground, and capped around the full circle with lintels weighing about 7 tonnes each. Within this circle five trilithons, up to 7.3m in height, were erected in a horseshoe pattern, each trilithon consisting of two enormous uprights, weighing over 40 tons, capped with a massive lintel. Subtle features of the construction included mortise and tenon joints between the uprights and lintels, and the outward tapering of the uprights and lintels to give an illusion of verticality. The final phase of construction consisted of re-erecting the bluestones, twenty of them, to form an inner horseshoe within the sarsens and the remaining stones forming a circle between the sarsen horseshoe and sarsen circle.

Raising the sarsens to their upright positions constituted no less a task than hauling them to the site from Marlborough Downs. Two possible methods have been discussed in the previous chapter in relation to raising the stones at Avebury. Interestingly, many Neolithic and Bronze Age communities displayed the ability and motivation to quarry, transport and raise huge megaliths. As early as 9000 BC the builders of Gölbekli Tepe in Turkey, close to the Syrian border, constructed circular enclosures bounded by stone walls intersected at intervals by vertical limestone slabs or pillars, quarried nearby, up to 7m high and weighing up to 50 tonnes.

Writing in 1924, Stone showed by using models the possibility of using the rolling technique to raise the Stonehenge lintels into position on top of the upright sarsens. The writer has also demonstrated this possibility using models, but rolling the blocks up inclined timber poles instead of earthen ramps as proposed by Stone. Inclined timber logs were employed in this way by Pavel in 1992 to raise a 5-tonne lintel, using levers to slide the lintel up the logs.

Herodotus makes many references to temples and their associated gods or goddesses, and in some cases gives some details of their construction, but, surprisingly, the Egyptian temple to Amon at Karnak and the Parthenon in Athens, dedicated to Athena, do not feature in his writings. Having travelled as far south in Egypt as Elephantine, presumably by river, he could not have failed to see Karnak, then, as now, the largest temple in the world, and he would surely have spent a day or two visiting it. Although omitting any description of Karnak, he was clearly impressed by other temples he saw or heard about in Egypt. His giving of descriptions of the temples at Bubastis and Sais, but not at Karnak, probably reflects, partly at least, that for much of the Late Period political power in Egypt resided in the Delta, because of its proximity to Egypt's Asian neighbours, with whom the country had become much more involved, and also the ready access to the Mediterranean, particularly Greece.

Both cities served as the country's capital in the Late Period, Bubastis sharing this distinction with Thebes during the overlapping Twenty-second and Twenty-third (Libyan) Dynasties (925–712 BC), and Sais during the Twenty-sixth Dynasty (664–525 BC), when powerful native Egyptian pharaohs once again ruled the country for the last time. Although at the time of Herodotus' visit to Egypt in the fifth century BC the Persians had re-established the capital at Memphis, Bubastis and Sais remained important cities in their own right, both staging important sacred festivals during the year. It is likely that he spent some time in both cities. According to Herodotus, during the Festival of Lamps held in Sais, everyone burned a great number of lights in the open air around the houses, and even those unable to visit the city lit lamps in other parts of the country. He also describes at length an important festival in honour of Artemis, held annually at Bubastis, attracting up to 700,000 adult participants.

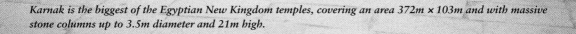

Karnak is the biggest of the Egyptian New Kingdom temples, covering an area 372m × 103m and with massive stone columns up to 3.5m diameter and 21m high.

He was obviously impressed by the temples he saw in both cities, as he wrote:

When an Egyptian committed a crime, it was not the custom of Sabacos [Shabaka, 725–667 BC] to punish him with death; but instead of the death-penalty he compelled the offender, according to the seriousness of the offence, to raise the level of the soil in the neighbourhood of his native town. In this way the cities came to stand even higher than they did before; for the level of the ground had already been raised once, in Sesostris' reign, when the canals were dug. This was the second time, and the result was that the cities came to stand very high indeed. None of the Egyptian cities, I think, was raised so much as Bubastis, where there is a temple of Bubastis (the Greek Artemis), which is well worth describing. Other temples may be larger, or have cost more to build, but none is a greater pleasure to look at. The site of the building is almost an island, for two canals have been led from the Nile and sweep around it, one on each side, as far as the entrance, where they stop short without meeting; each canal is a hundred feet wide and shaded with trees. The gateway is sixty feet high and is decorated with remarkable carved figures some nine feet in height. The temple stands in the centre of the city, and, since the level of the building everywhere else has been raised, but the temple itself allowed to remain in its original position, the result is that one can look down and get a fine view of it from all round. It is surrounded by a low wall with carved figures, and within the enclosure stands a grove of very tall trees about the actual shrine, which is large and contains the statue of the goddess. The whole enclosure is a furlong square. The entrance to it is approached by a stone-paved road about four hundred feet wide, running eastward through the market-place and joining the temple of Bubastis to the temple of Hermes. The road is lined on both sides with immense trees – so tall that they seem to touch the sky.

His [Amasis'] first work was the marvellous gateway for the temple of Athena in Sais. He left everyone else far behind him by the size and height of this building, and by the size and quality of the blocks of stone of which it was constructed. He then presented to the temple some large statues and immense men-sphinxes, and brought for its repair other enormous blocks of stone, some from the quarries near Memphis, and the biggest of all from Elephantine, which is twenty days' voyage by river from Sais. But what caused me more astonishment than any thing else was a room hollowed from a single block of stone; this block also came from Elephantine, and took three years to bring to Sais, two thousand men, all of the pilot-class, having the task of conveying it. The outside measurements of this chamber are: length, 21 cubits; breadth, 14 cubits; height, 8 cubits. Inside, its length is 18⅔ cubits; its breadth 12, and its height 5. It lies by the entrance to the temple, for the story goes that it was not dragged inside the enclosure, because, during the process of moving it, the man chiefly responsible for its construction groaned with vexation at the amount of time and labour which was being wasted, and Amasis took this for a bad omen and refused to have it brought any further. Another story is that it was left outside, because one of the workmen who were levering it along with crowbars got crushed to death underneath it.

Excavations at Tell Basta (Bubastis) towards the end of the nineteenth century showed the existence of a temple up to 300m in length, but revealed very little detail of its structure. The excavations confirmed Herodotus' description of the temple standing on an island at a level lower than that of the surrounding city, and having water channels on either side. Remains at Sa el-Hagar (Sais) today consist of no more than a few scattered stone blocks, although there are records of evidence in the nineteenth century of a large enclosure surrounded by mud-brick walls. All useful stone building material had already been long-since removed.

The temple of Amon at Karnak, surely seen by Herodotus but surprisingly not featuring in his writings, is not only the largest column and beam structure ever built, but is also the largest building ever constructed for religious purposes. Its construction took place in phases over many centuries. Around 2000 BC, when the first small shrine was built at Karnak, the great stone pyramids had already stood for over half a millennium. Despite rising prosperity, the ruling pharaohs of the Middle Kingdom encouraged restraint and favoured the construction of irrigation schemes rather than large and impressive buildings. The first shrine at Karnak was consequently a modest structure, despite its dedication to the leading god Amon-Ra.

The New Kingdom pharaohs, whose names such as Tuthmosis, Amenophis, Seti and Ramses have rung down through the ages, had no inhibitions about spectacular display, seeing in huge statues and immense buildings the glorification of both themselves and Amon-Ra. Karnak is only one of the huge columned temple complexes of the New Kingdom, but it is the biggest. Even now it has not been fully excavated. Every major pharaoh of the period, with the exception of the heretic Akhnaten, felt the need to add something to the temple complex, with the result that it ended up with nine pylons, huge trapezohedron structures, giving entrance to the main body of the temple. In addition to the temple of Amon, which alone boasts six huge entry pylons, the main complex, occupying an area of 25 hectares surrounded by a thick mud-brick wall, contains another major temple (the temple of Khons) and various subsidiary temples, a sacred lake and a number of obelisks. On the northern side of the main sanctuary a second mud-brick enclosure contains the temple of the Theban god Monthu and, to the south, connected to the main precinct by a short processional way, lies a third enclosure containing the temple of Mut, consort of Amon, with its own sacred lake. Further to the south, connected to the Karnak enclosure by a sphinx-lined avenue 2km long, is the temple of Luxor. During the feast of Opet a grand procession, embodying the triad of Amon, Mut and Khons, sailed by state barge along the short stretch of the Nile to Luxor, to be met by a ceremonial welcome and sumptuous repast.

The temple of Amon, housed within the main enclosure and alone occupying nearly four hectares, conformed to a common rectangular plan, albeit it much larger than most other temples, with a length of 372m and a width of 103m. The great Hypostyle Hall occupied the central portion of the rectangle, and was completed during the time of the Nineteenth Dynasty pharaoh Ramses II (1298–1232 BC), a man possessed of unbounded megalomania. When finished, it consisted of sixteen rows of massive columns running the full width of the temple, their papyrus bud capitals supporting massive 70-tonne stone lintels, which, in turn, supported a heavy slab-stone roof. The central two rows of twelve columns were each raised several metres higher than the rest, giving an elevated roof level over this portion of the hall, with the vertical separation of the two roof levels taken up by stone grids, which allowed light into what would otherwise have been a dark and gloomy interior. Each of the gargantuan columns in these two rows, which flanked the processional way through the temple, soared to 21m in height, with a diameter of 3.5m. Unlike the shorter columns, these higher columns had open papyrus capitals. Carved reliefs and inscriptions adorned the walls, and, along with columns and lintels, gave praise to the pharaohs who built the temple and the gods to whom it was dedicated.

An earthquake in 1899 toppled eleven of the columns of the Hypostyle Hall. A French Egyptologist, Georges Legrain, working on the site at the time, undertook their re-erection, and to do so adopted the methods he assumed to have been employed by the ancient Egyptians. He had earth ramps constructed and hauled the massive drums up them, replacing them into their original positions. Legrain also rebuilt one of the pylons that had collapsed. Earthquakes were not the only hazard that threatened the stability of structures sited close to the Nile. Murray attributes the shape of the pylons to the need to provide a wide base to counteract ground movements caused by fluctuating groundwater levels induced by the rise and fall of the Nile.

One end of the temple of Amon comprised a colonnaded court and the other end a sanctuary containing an image of the god Amon, surrounded by various rooms for storage and usage by the priests. The enormous wealth of the Theban temples can be gauged from their possessions that, at their height, included 90,000 slaves, 500,000 head of cattle, 400 orchards, 80 ships and 50 workshops. They also received revenues from sixty-five townships in Egypt and Asia Minor.

Greek temple builders showed the same unswerving attachment to colonnaded structures as the Egyptians of the New Kingdom had some 500–1,000 years before them; but they eschewed sheer enormity of structure, which characterised the Egyptian creations, for careful site selection to show off the temples to greatest advantage, and purity of form. Correct and pleasing proportions mattered above all else. Greek temple construction reached its apogee with the Parthenon, both in its spectacular setting on the 75m-high Acropolis dominating the topography of Athens and in the perfection of its structural proportions fashioned from the fine white marble of Mount Pentelikon.

The colonnaded Greek temples familiar to travellers in Mediterranean countries today evolved from early Greek temples, which comprised essentially a room or chamber with an entrance porch, constructed of mud-brick sometimes reinforced with a timber frame. To protect the structure from weather erosion, a thatched roof on a timber frame extended beyond the walls. During the eighth to sixth centuries BC terracotta tiles and stone replaced thatch for important temples, and the extension of the heavy roof beyond the chamber required a row of stone columns around the perimeter of the building for its support. Finally, stone replaced mud-brick in the construction of the chamber, which, in a typical Greek temple at the time of Herodotus, had evolved into a rectangular walled sanctuary, divided longitudinally into three rooms. The most sacred, and largest, room, the *cella* (or *naos*), containing the statue of the deity, was entered through a smaller antechamber, or *pronaos*, which faced to the east. Attached to the western end of the *cella*, but with no access to it, was a room known as the *opisthodomus*, corresponding in size to the *pronaos*. Although typical, this layout was not followed slavishly.

Herodotus would have known two of the largest temples of his time well, but makes only passing reference to them. The temple of Artemis at Ephesus, built around 560 BC by

The classic fifth-century BC *Greek Doric temple at Paestum in Italy had thirty-six exterior columns rising to 8.5m, with a diameter of 2m at the base and 1.4m at the capital. Two rows of seven columns supported the roof of the* cella. *Built of travertine, it had a white stucco overlay and brilliant polychrome decoration. (From Hammerton)*

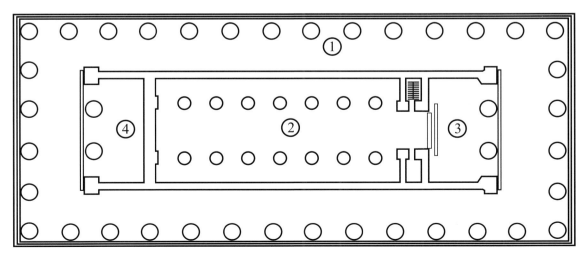

Typical of classical Doric, the temple of Hera II at Paestum has an emblature (1) with plan dimensions of width to length corresponding close to 4 : 9. The main chamber (2) is accessed through the pronaos (3), while the opisthodomus (4) has no access to it. Two rows of columns divide the main chamber into three aisles.

The fourth-century BC classic Ionic temple to Artemis (Diana) at Ephesus, which Pliny marvelled at, was modelled closely on its sixth-century BC predecessor, financed by King Croesus, but burnt down by an arsonist in 356 BC. The relief carvings on the bases of the slender columns were an unusual feature for a Greek temple. (From Hammerton)

the Ionian Greeks with the financial support of Croesus, king of neighbouring Lydia, is the only temple to make it into the Seven Wonders of the Ancient World. It is referred to in the Bible (Acts 19) as the temple of Diana of the Ephesians. Herodotus speaks of the temple to Hera at Samos, an island well known to him, in the same breath as that at Ephesus, but compares both unfavourably in size and grandness to the Labyrinth near Lake Moeris in the Egyptian Fayum.

Built on poor marshy ground, reputedly founded on a raft of alternating charcoal and fleece layers, the Artemision of Ephesus had plan dimensions of 115m by 51m, unsurpassed in size by any other Greek temple. Unusually, its entrance faced to the west, rather than to the east, perhaps reflecting a strong Anatolian influence in its creation. Its 129 columns comprised two rows along each side and at the east end, and multiple rows of columns fronting the entrance, some of the latter having, unusually, bases with carved relief decorations. With columns 20m high, equal to a modern six-storey building, it must have been a breathtaking sight. The lintels making up the architrave weighed up to 40 tonnes, with lengths up to 8.75m, of which 6.5m spanned unsupported between columns, pushing this form of construction to the limit.

The heavy marble architrave lintels and column drums had to be transported to the site from quarries 11km distant. Vitruvius, writing in the first century BC, has an amusing, if unlikely, story relating to the discovery of this marble outcrop. The shepherd, Pixodorus, was feeding his flock, when two rams ran at each other, but failed to make contact and one of them, instead, struck his horns against the rock outcrop and dislodged a fragment of beautiful white marble. Pixodorus knew the powers that be were seeking a suitable stone for their new temple so he ran to Ephesus to show them the fragment. They showed their gratitude by rewarding him with honours. Whatever the truth of this story, there can be no doubting the description by Vitruvius of how the engineer/architect Chersiphron, and his son Metagenes, moved the heavy blocks from the quarries to the temple site:

> It may also not be out of place to explain the ingenious procedure of Chersiphron. Desiring to convey the shafts [column drums] for the temple of Diana at Ephesus from the stone quarries, and not trusting to carts, lest their wheels should be engulfed, on account of the great weights of the load and the softness of the roads in the plain, he tried the following plan. Using four-inch timbers, he joined two of them, each as long as the shaft, with two cross-pieces set between them, dovetailing all together, and then leaded iron gudgeons shaped like dovetails into the ends of the shafts, as dowels are leaded, and in the woodwork he fixed rings to contain the pivots, and fastened wooden cheeks to the ends. The pivots, being enclosed in the rings, turned freely. So, when yokes of oxen began to draw the four-inch frame, they made the shaft revolve constantly, turning it by means of the pivots and the rings.
>
> When they had thus transported all the shafts, and it became necessary to transport the architraves, Chersiphron's son Metagenes extended the same principle from the transportation of the shafts to the bringing down of the architraves. He made wheels, each about twelve feet in diameter, and enclosed the ends of the architraves in the wheels. In the ends he fixed pivots and wheels in the same way. So when the four-inch frames were drawn by oxen, the wheels turned on the pivots enclosed in the rings, and the architraves, which were enclosed like axles in the wheels, soon reached the building in the same way as the shafts.

Rolling heavy objects is a far more efficient means of moving them than dragging them on sledges, and may well have been a method commonly resorted to in ancient times. Another case cited by Vitruvius concerns one Paconius, who lived around Vitruvius' own time, and who was contracted to cut a 40-tonne stone block in the same Ephesus quarries and transport it to Ephesus to replace the cracked pedestal supporting a large statue of the god. He enclosed each end of the block in circular timber wheels and connected the rims of these with closely spaced timber bars, effectively encapsulating the block in a wooden cylinder, which could be rolled. According to Vitruvius, 'Then he (Paconius) coiled a rope round the bars, yoked up his oxen, and began to draw on the rope.' Paconius failed in his attempt as the stone swerved off course, probably as a result of the short length of the block compared to the rolling diameter, and of using only a single rope, which would have given him little control over its direction, exacerbated by having the rope, rather than the timber bars, in contact with the ground. As described elsewhere in this book, this

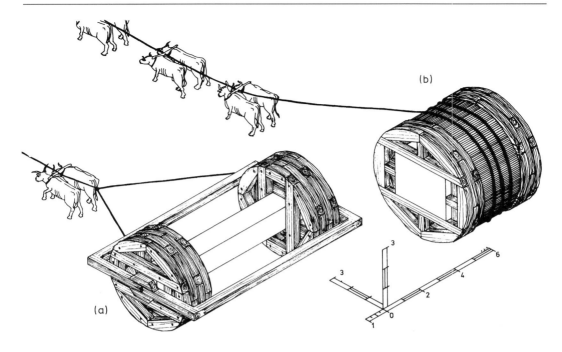

*According to Vitruvius, the engineer Metagenes transported heavy architraves from the quarry to the sixth-
century BC Temple of Diana at Ephesus by enclosing the ends in 'wheels' with central pivots and rolling them (a).
He also describes how his contemporary, Paconius, attempted to enclose a 40-tonne pedestal in a cylinder of
timbering and roll it using oxen pulling on rope coiled around it (b). (From Coulton)*

technique is likely to have been used in transporting and raising the stone blocks for the
great stone pyramids in Egypt.

The elder Pliny, who visited the temple at Ephesus in the first century AD, no doubt knew
of Chersiphon's method of transporting the blocks, but considered his raising of the
architrave blocks into position to be his greatest feat. According to Pliny, he constructed a
gently inclined ramp with reed bags filled with sand up to and above the level of the
capitals. Pliny does not say how Chersiphron raised the blocks up the ramp, but it could
well have been by a combination of levering and rolling the blocks. Once at the top,
according to Pliny, he gradually emptied the bottom sacks, allowing the architrave block to
settle gently into place. A problem arose while attempting to place the heaviest block over
the doorway, causing Chersiphron to contemplate suicide, but this was thankfully averted
by the intervention of the goddess Diana, who placed the block overnight while he slept. He
went on, with Metagenes, to write a book on the design of the temple, to which Vitruvius
and other ancient writers had access, but which has not survived to the present day.

In AD 356, supposedly on the night of Alexander the Great's birth, an arsonist called
Herostratus set fire to the roof and destroyed the temple, and so achieved his aim of
immortality. Seeing the building still under reconstruction when he visited the site at the
age of twenty-two, Alexander offered help in paying for its restoration, but the Ephesians
diplomatically declined on the basis that one god should not build a temple to another.

They rebuilt it themselves, again on a massive scale. Little remains today, other than a single re-erected column with a stork's nest on top, surrounded by evidence of archaeological investigations, not least by John Turtle Wood, a British railwayman turned archaeologist, who excavated the site between 1863 and 1874. Earthquakes and the destructive hand of man eventually put paid to the building and, like so many ancient buildings, it became a quarry site for various later construction and reconstruction works in Ephesus, including churches and aqueducts.

Both the temple to Artemis at Ephesus and the one to Hera in Samos, at the time of Herodotus, occupied sites on which there had been earlier temples, the earliest dating back to perhaps the tenth century BC and constructed of timber and mud-brick. Their relative proximity to each other also ensured a constant rivalry, so that the temples Herodotus knew were two of the biggest Greek temples ever built. The first major stone temple on Samos, the Heraion, completed by the engineer/architects Rhoikos and Theodorus in about 560 BC, measured 102m by 51m in plan and had 104 columns arranged in double rows along each side and front and back. The sacred inner rooms consisted of a *pronaos* and *cella* only, with two rows of internal columns. This temple had only a short life before collapsing, possibly as a result of fire, but more likely as a result of the foundations settling on the soft marshy soil on which it was built. This presented Policrates with the opportunity to build an even larger structure entirely in marble, 108m by 52.5m in plan, with 133 columns up to 2m in diameter, comprising double rows of columns along each side and within the *pronaos* and triple rows of columns front and back. All that can be seen now are some nondescript remains and a column, half its original height. The absence of roof tiles raises the question of whether or not it was ever finished.

In 480 BC, when the Persian forces under Xerxes poured into Athens almost unopposed, they heaped humiliation on the Greeks by setting fire to the sacred temple, of timber or partly timber construction, which then occupied the Acropolis site. Perhaps this very shame, inflicted upon them by Xerxes, stirred the Greeks to fight back, against all the odds, under the Athenian commander, Thermistocles, and to defeat the Persian fleet (depleted by storms and by the ships used to build the pontoon bridge across the Hellespont) and expel the Persians from the mainland. Although the war against Persia lasted for another thirty years, Athens prospered, in part, because money sent by their allies in the Delian League for ships and arms to fight the Persians found its way into the Athenian treasury. It was supplemented in no small measure by the silver won from the Laurion mines, where thousands of slaves toiled under conditions little better than a living death.

Inspiration for the great works on the Acropolis came from Pericles, ruler of Athens (460–429 BC) and one of history's great statesmen. Dismissing the protests of his Delian allies on the grounds that he was fulfilling his responsibility of protecting them, he spent the money in his treasury lavishly on temples, shrines and statues, employing on these works the leading architect/engineers and artists of the day. It was truly a golden age for Athens. For the works on the Acropolis he appointed the architect/engineer Liktinos, the contractor Callicrates and the sculptor Phidias. Other great works of the time included the colonnaded administrative buildings and shops of the Agora or market place at the foot of the Acropolis, where such as Socrates and Diogenes gave vent to their philosophical thoughts, the stadium at Olympia and the arena at Delphi.

In the Parthenon – a Doric temple – the enclosed sanctuary consisted of only two rooms, the larger room, the *cella*, having the dimensions 19m by 29m in plan. It contained an enormous statue of the virgin Athena, 12m high including its pedestal, carved by Phidias from wood, but with face, hands and feet of ivory, eyes of precious stones and various gold accessories. The second, smaller, room served as a treasury. Another statue of Athena by Phidias, 9m high in bronze, stood in a prominent position outside the temple. Other statues also occupied the sacred area.

Worshippers gained access to the site by a path winding up the hillside, lined with statues, and at the top passed through the Propylaea, a huge pillared gateway. Apparently having some reservation about the strength of stone architraves, the engineer/architect of this structure, Mnesicles, used 2m-long iron beams to transmit loads of 64 tonnes from the marble ceiling beams to the columns on either side. Once through this gateway the Parthenon, standing on its stone platform, came into view on the right, and to the centre-left stood the oldest temple, the Erectheion, named after the legendary king of Athens. Phidias' bronze statue of Athena dominated the view ahead.

The Parthenon, on the Athens Acropolis, dates from the Golden Age of Athens in the fifth century, and Herodotus may well have witnessed its construction. The cella *contained a 12m-high statue of Athena. The structure suffered extensive damage from a Venetian mortar shell in 1687.*

Standing on a solid stone platform or *stylobate*, the Parthenon conformed to the proportions considered to be good building practice at the time, in particular making use of the ratio 4:9. The overall width of 31m to length 70m conformed to this ratio, as did the height (columns plus entablature) to width and the relation of column base to column spacing. The marble columns were approximately 1.8m in diameter and 10.4m high, made up of fluted drums varying from 0.73 to 0.99m in height, positioned by wooden or iron dowels which fitted into holes bored into wooden inserts in the horizontal faces of the drums. The horizontal surfaces appear to have been prepared in four concentric circles: an outer fine finished surface to give a close-fitting weatherproof joint, the next rougher to give improved resistance to lateral movement, the third recessed to reduce contact area and thus increasing pressure between contacting outer zones, and finally a small central zone again at joint level. Sand was used as an abrasive to grind the fine surfaces together. Simple Doric capitals adorn the tops of the Parthenon columns.

Stone blocks and lintels were held together at the joints, not with mortar, but with metal cramps and dowels. Usually of iron, sometimes cased in bronze, these 'T'-shaped cramps fitted into correspondingly shaped slots sunk into adjoining blocks, with molten lead poured in to hold them in place. Iron cramps of this type have proved to be one of the most active agents causing the deterioration of these types of structures, the expansion of the iron when it rusts causing cracking of the masonry.

The Greeks introduced a number of subtle refinements into the structure of the Parthenon to compensate for optical illusions created by the eye. The *stylobate* has a convex parabolic curvature with rises above the corner levels of 60mm and 130mm, respectively, at the mid-points of the ends and the sides. This counters the illusion of sagging, which a flat platform would have given. All columns lean slightly inwards and the corner columns, because they are seen from a distance against the open sky, are slightly larger in diameter and are spaced a little closer to their neighbours. To enhance the appearance of strength all columns bulge slightly in the middle as well as having an upwards taper.

The Parthenon was only one, albeit the best, of similar Greek temples built in the comparatively short period between the defeat of the Persians in 480 BC and the death of Alexander of Macedon in 323 BC. Remains of these, some in a better state of repair than the Parthenon, can be seen not only in Greece, but in other parts of the Mediterranean which the Greeks colonised, particularly in Sicily, southern Italy and Asia Minor. In the fifth century AD, after nearly a thousand years as a temple to Athena, the Parthenon, under Byzantine influence, became a Christian church, with some consequent changes in the structure and removal of the statues of the goddess. The Ottomans in their turn converted it into a mosque. It was still in a fair state of repair in 1687 when a Venetian army besieged the city, in the course of which they lobbed a mortar shell through the roof of the Parthenon. Regrettably, the Turks, with apparently little respect for the structure despite its having been converted into a mosque, had used it to store gunpowder, which blew up causing extensive damage. It is now believed, however, that damage caused to the structure by an ancient fire had resulted in cracking and fracturing of the masonry, so reducing its ability to resist the explosion of 1687. It suffered further damage from a strong earthquake in 1894. Detailed restoration of the structure has now been in hand for many years.

The Pantheon.

In 1801 Lord Elgin, with the permission of the Turkish government, removed most of the remaining sculptures and shipped them to England. Any pangs of conscience he may have felt were assuaged by the fact that the structure had suffered from grievous neglect. The sculptures remain in London to this day despite Greek requests for their return. They were in fact lucky to reach London at all. The ship transporting the 'Elgin Marbles' ran into a rock off Kythera and sank. Not to be denied, Elgin hired divers to recover them and eventually they reached London, where he sold them to the British Museum.

In temple construction, for the most part, the Romans carried on with the Greek forms; and many of the construction methods, rules and proportions given by Vitruvius come from the writings of Greek engineer/architects, which he had available to him. But for the main structure of the Pantheon in Rome, the Romans departed completely from the column and lintel form of construction commonly adopted for other temples in the ancient world. It remains, even today, one of the largest domed structures ever built: its construction made possible by the Roman invention of concrete, a material not only of great strength, but able to be moulded into a variety of shapes.

Although the early history of the Pantheon in Rome remains uncertain, it is likely that credit must go to that greatest of all emperor/builders, Hadrian, for commissioning the structure seen today. Not only is it one of the most remarkable structures ever built, but it is no less amazing that it has survived structurally intact: the only major Roman building to have done so. Even if Hadrian cannot be credited with the present building in its entirety, he certainly had it extensively rebuilt around AD 120, the original temple on the site having suffered grievously in the previous century, after its construction by Agrippa, mainly from Nero's fire in AD 64 and later by lightning strike. Not surprisingly, much has happened to the building since Hadrian's reconstruction, including the need for repairs during the times of Septimus Severus and Caracalla. Its conversion into a Christian church in AD 608 did not stop the Christian Byzantine emperor, Constans II, from stripping the gold-plated brass plates from its roof during a brief sojourn he made in Italy in AD 661 to escape Arab attacks on Constantinople and internal dissent within that city. But any material gain he made from this was short-lived, as Arab pilots caught and killed him on his way back to Constantinople. Quarrying of the marble facing covering the exterior of the rotunda occurred in the Middle Ages, the final major abuse coming in 1625 when Pope Urban VIII removed the bronze girders holding up the roof of the portico and replaced them with timber. He had the bronze melted down to build eight cannons which he set up, ironically, around the tomb of Hadrian, which had been converted into the fortress of Castel Sant' Angelo by earlier popes.

It would have been better if Pope Urban VIII had removed the whole of the portico, never to be replaced, as its classical style is in complete disharmony with the domed building, and in a considerable measure detracts from it. Once one passes through the columned portico and enters the rotunda, the rare and magnificent site of one of greatest masonry domes ever built greets the eye. The powerful images of the recessed coffers in the dome's interior surface, deliberately built to diminish in size with height to suit the geometry and enhance the perspective, lead the eye ever upwards to the 9m-diameter opening (oculus) at the crown, 45m above floor level, giving unfettered access to the seven planetary deities to whom the building was dedicated. It also allows light into the interior.

The perfectly hemispherical dome is supported on a massive concrete drum or rotunda 43.3m in diameter and with a wall thickness of 6m, the interior of which is faced with precious marble. In the lower half the interior wall is broken up by alternating round and square pillared niches, seven in number, one for each of the deities. An eighth recess forms the entrance to the rotunda. Above the half-height level of the drum, concealed brick relieving arches transfer the weight and thrusts of the dome across the hollowed niches into the thick intervening sections of wall.

The exact method of constructing the dome is unknown, but the coffers sunk into it make it in effect a ribbed structure, although the depths of the ribs are relatively small compared with the total thickness of concrete. In order to reduce the overall weight and thrusts of the dome, the rubble used as aggregate in the concrete changes from *tufa* and travertine at its base to *tufa* and brick with increasing height and finally to pumice at the highest levels. While the dome is hemispherical within, its shape seen from the outside is more that of an inverted flat bowl. This is because the upper part of the cylindrical wall of the supporting rotunda rises above the interior springing level, to form a circular exterior terrace, and above this level the exterior of the dome rises in a series of concentric step-like rings with vertical faces, which add to the thickness of the shell to the extent that hoop tensile stresses in the concrete, set up by the horizontal thrusts of the dome, probably do

The unprepossessing exterior of the Pantheon in Rome, built by order of Hadrian around AD 120, hides a remarkable 43.3m-diameter coffered concrete dome that takes the breath away. The use of pumice in the aggregate reduced the weight near the top of the dome. (From Hammerton)

not exceed 0.5 MN/m², which a fully hardened crack-free concrete can easily tolerate. In fact, in common with all masonry domes, cracking has occurred, rising from springing level towards the oculus, suggesting that the thrust is being absorbed by the thick rotunda wall acting as a buttress. The extended height of the wall and thickening of the base of the dome increase the verticality of the resultant thrust in the wall and are thus beneficial.

The steps forming the lower portion of the dome probably resulted from the construction process. It is likely that the concrete for the dome was placed in horizontal layers, the steepness of the external surface at the lower levels requiring the concrete to be retained by a vertical formwork in a series of lifts with decreasing radius. Formwork would also have been required to form the interior surface, but as the height of construction increased the lower portions already placed would have hardened and become self-supporting, allowing the internal formwork to be raised to a higher level. Close-sheeted formwork, covering the whole of the internal surface of the dome, would consequently not have been necessary. Common to much Roman construction, brickwork conceals much of the exterior concrete, serving both as formwork and as a decorative veneer. Brick arches embedded into the rotunda wall, and visible on the exterior surface, serve to carry loads away from the interior niches towards the thicker sections of the wall, which, in effect, form eight substantial piers. Their primary purpose would have been to distribute the loads during construction, as they would have ceased to play any important role once the concrete hardened.

While simple timber post and beam structures, often combined with other materials for walls or roofing, date back many thousands of years in different parts of the world, the masters of this type of timber building were, and still are, the Chinese and Japanese. In some measure, at least, this may be attributed to the ever-present threat of disastrous earthquakes in both countries and the ready availability of the one building material, prior to modern times, able to resist strong seismically induced shocks. But timber has the disadvantage that it deteriorates with time and, although a number of Chinese and Japanese temples are well over a thousand years old, their timbers have probably been replaced a number of times, while retaining the original form of the construction.

Indian missionary monks first brought Buddhism to China in the first century AD, some 400 years after the time of its Indian founder Gautama. Initially it spread only slowly, partly because of the difficulty of translating Sanskrit texts into Chinese, in addition to which it had to overcome opposition from Confucian scholars and competition, particularly among the peasant classes, from the mystical beliefs of Taoism. However, by the time the short-lived Sui Dynasty (AD 581–618) had reunited the empire, only to lose it to their illustrious Tang successors (AD 618–907), Buddhism had become well established throughout China, although it assumed its own Chinese character, without displacing Taoism or threatening Confucianism.

As pointed out by Needham, no Chinese building could be a proper dwelling for the living, or a proper place of worship for the gods, unless built of wood and roofed with tile. He summarises the main features of Chinese building (secular and religious) as: (a) emphasis on the roof, and its construction in sweeping curves; (b) formal grouping of buildings around rectangular courts, and marked attention to axis; (c) frankness of construction, the supporting pillars for the roof timbering being clearly visible, even when partly incorporated with the walls; (d) a lavish use of colour, not only in roof tiles, but on painted columns, lintels and beams, richly bracketed cornices and broad expanses of plastered walls. Large timber buildings such as temples were normally situated on a base of hard rammed earth, sometimes faced with stone or brick, with the timber columns, usually of pine or a variety of cedar, mounted on stone bases and tied in at their tops by a system of orthogonal horizontal beams. Decorative cantilever brackets fixed to the tops of the columns supported the ends of the beams.

The simplest structure consisted, in cross-section, of a bay made up of two columns, with their tops connected by a beam, which, in turn, supported subsidiary beams. Longitudinal purlins, usually round in cross-section, rested on seatings near the end of each beam, while king posts, rising from a subsidiary beam, supported the roof ridge purlin. This system differed greatly from later western-style roof timbering, such as triangulated trusses or complicated systems of hammer beams and arch timbers; one advantage of the Chinese system was its modular form, which allowed the bays to be extended to any required width by adding more columns and more tiers of beams. The curved roof profile favoured by the Chinese could easily be achieved by the choice of beam lengths. A disadvantage compared to a timber truss or arch system was the limitation placed on modular widths by the bending capacity of the beams, which was more restrictive than the use of timber members in simple compression or tension.

The structure of the roof consisted of round rafters, typically bamboo, supported by the purlins and, in turn, supporting boarding, covered by an insulating layer of clay, on which rested two layers of segmental tiles. Buildings could be extended lengthwise simply by adding more bays, with the whole structure closed and partitioned by fixing light timber panelling between columns. As the roof structure played a dominant role in the visual design of the building, ceilings were not usually provided except sometimes as a decorative feature suspended from one level of cross-beams. Although timber structures have not survived from the Han period, clay models made at the time show pitched roofs with linear profile and wide eaves intended, no doubt, to protect the earth or timber walls. The reason for introducing curved roofs after the Han period invites speculation and, although aesthetically pleasing, may have been adopted for more pragmatic reasons. Needham lists a number of possibilities, including the observation that the upturned eaves would admit a maximum amount of slanting sunlight in winter and a minimum amount of down-pouring summer sunlight. Other possible reasons he cites are reduced wind pressure and the imparting of a shooting action to snow and rain falling on the roof, to carry them well clear of the raised platform supporting the building.

Buddhism arrived in Japan from China in the middle of the sixth century BC. Again, as in China, it did not supplant existing beliefs – in this case Shintoism – and indeed complemented this indigenous faith, so that most Japanese even today are both Shintoists and Buddhists. Even before the arrival of Buddhism the Japanese had learnt the art of timber construction, probably from the Chinese through mutual Korean contacts, and applied it in the building of their Shinto temples. Buddhist temples assumed much the same structural form as the Chinese temples, reflecting the fact that many of the early temples were built by Chinese craftsmen invited to settle in Japan. One of the most important temples in Japanese history, art and culture, the Horyu-ji temple in Nara, was started in AD 607 under the direction of Prince Shotoku. The temple actually consists of several timber buildings including shrines, halls for prayers and teaching, and a five-roofed sacred tower. The curved tile roofs and bracketed pillars are typically Chinese, but the basic structures are lighter and more delicate than their Chinese counterparts, notwithstanding which some parts of the original structures still stand. Nara also boasts the largest timber building under one roof in the world. It is the Daibutsuden, a huge *cella* or hall nearly 50m high, 57m long and 50m wide, which houses the Great Buddha, the largest bronze statue on earth. The statue was completed in AD 749 and weighs 450 tons.

8

Buildings for Kings and Subjects

From 2500 BC, or perhaps earlier, to 1500 BC, when overrun by Aryans from the north, a civilisation flourished along the Indus Valley matching, and in contact with, that of Mesopotamia. Its two major cities – Harappa in the north-east, on a tributary of the Indus, and Mohenjodaro, lower down on the Indus itself – were separated by some 600km, but numerous other settlements extended the influence of these cities to an area of more than 1,000km in length. Each of the two major cities had a strongly walled citadel, built on an artificial hill, with processional terraces, monumental gateways and large public buildings, below which lay the town itself, a model of town planning, occupying an area about 1.5km square and separated into blocks by wide main streets at right angles. Within each block individual close-set houses fronted onto interior courtyards, presenting blank outer walls to the streets and lanes accessing them. Waste-water from the houses discharged into arched brick drains running below the unpaved streets, while other rubbish discharged into brick bins through chutes passing through the walls of the houses. Most of the houses had stairways leading to a second storey or to a flat roof providing a cool sleeping place in summer.

All public buildings and many of the houses were built of fired bricks, produced in vast numbers to a standard size of 279mm by 134mm by 63.5mm. Regrettably these high-quality bricks attracted the attention of the nineteenth-century railway builders, several million of them having been carted off to serve as ballast in constructing the East Indian Railway from Lahore to Multan. A surprising feature for such an advanced civilisation is the absence of grandiose structures such as palaces, forts or great temples, suggesting a highly egalitarian or even democratic society. Conspicuous structures have, however, been excavated within the citadel site at Mohenjodaro, including the Great Bath, a Granary and a Pillared Hall consisting of twenty pillars around a small courtyard, which may have been an administration centre.

The Great Bath was a rectangular structure 12m long, 7m wide and nearly 2m deep, made watertight by an inner facing of fired bricks set on edge in gypsum mortar and backed by a 25mm-thick layer of asphalt laid against double brick walls. The floor was made watertight in a similar manner. A corbelled brick drain at one end of the sloping floor served to drain the pool. Although the function of the Great Bath is unknown, it is thought to have been used for ritualistic purposes, unlike Roman public baths, which served much more hedonistic activities. The Granary, identified from charred grains of

The Great Bath at Mohenjodaro comprised a pool 12m long and nearly 2m deep, the walls and base made watertight by inner facings and flooring of baked brick set in gypsum mortar with an underlying layer of asphalt 25mm thick. The sloping floor led the water to an outlet that accessed a corbelled arch drain.

wheat found associated with it, occupied a massive podium made up of fired brick platforms separated by a system of straight, narrow passages, supporting a floor and building of timber.

The important town of Shedet (also known as Crocodopolis) was established in the Egyptian Fayum during the Old Kingdom period, and was probably built on silts brought down by the Bahr Yusuf. It became a cult centre of the crocodile god, Sobek, and an important place of pilgrimage. According to Herodotus there existed near this 'City of Crocodiles' a remarkable building called the Labyrinth, surpassing even the pyramids:

It has twelve covered courts – six in a row facing north, six south – the gates of the one range exactly fronting the gates of the other, with a continuous wall round the outside of the whole. Inside, the building is of two-storeys and contains three thousand rooms, of which half are underground, and the other half directly above them . . . the baffling and intricate passages from room to room and from court to court were an endless wonder to me, as we passed from a courtyard into rooms, from rooms into galleries, from galleries into more rooms, and thence into yet more courtyards. The roof of every chamber, courtyard and gallery is, like the walls, of stone. The walls are covered with carved figures, and each court is exquisitely built of white marble and surrounded by a colonnade.

Strabo, writing 400 years later, gives an equally expansive description of the building, but restricts it to one storey.

The detailed description given by both writers is reason to believe they both saw this building, despite claims by some modern writers that it never existed, except in the minds of Greek writers. Accepting that it did exist, the argument then turns to its purpose. The location of this building is usually considered to be adjacent to the Hawara mud-brick pyramid, built for Amenemhet III, taken to be the pyramid referred to by Herodotus and Strabo. Unfortunately there are no remains, other than a few masonry fragments with occasional crocodile reliefs, and some isolated bits of foundations, rendering impossible any thorough archaeological identification or reconstruction. Petrie blamed railway engineers for much of the quarrying. Using the account by Herodotus and the fuller account by Strabo, some reconstructions have been attempted. And Petrie concluded, based on literary and flimsy site evidence from his excavations in 1889, that an edifice some 1,000ft (300m) long and 800ft (250m) had occupied the site.

Although, like most major structures in pharaonic Egypt, the Labyrinth may have had some religious significance, the descriptions by Herodotus and Strabo make it unlike any other temple in the country, despite some leading Egyptologists believing this to be its purpose. Herodotus refers to underground rooms containing the bodies of the kings who built the Labyrinth and the bodies of sacred crocodiles. He was not permitted to see these and it is questionable as to whether they actually existed, which may be why he wasn't shown them. There is no site evidence of underground structures and Strabo makes no

An important structure occupied this site adjacent to the Hawara Pyramid, but to what extent it corresponded to a Labyrinth as described by Herodotus is conjectural. Images on surviving fragments of masonry attest to the importance of the local cult of the crocodile god, Sobek.

mention of tombs, other than of the interment of Imandes, presumably his name for the king responsible for the Labyrinth, below the pyramid at one end of the complex. He mentions crypts in front of the entrances, but, from his description, these seem to be on the same level as the courts, and may simply have been storage rooms, which would not be surprising as this building, whatever its function, was situated within the most agriculturally productive area in the country. Perhaps the most plausible suggestion is that by W.C. Hayes of 'a great architectural complex' including within it a palace, mortuary temple and administration centre (and perhaps law courts, as claimed by Strabo). Hayes compares it with the walled pyramid city for the workforce like that for the pyramid of Sesostris II at Illahun.

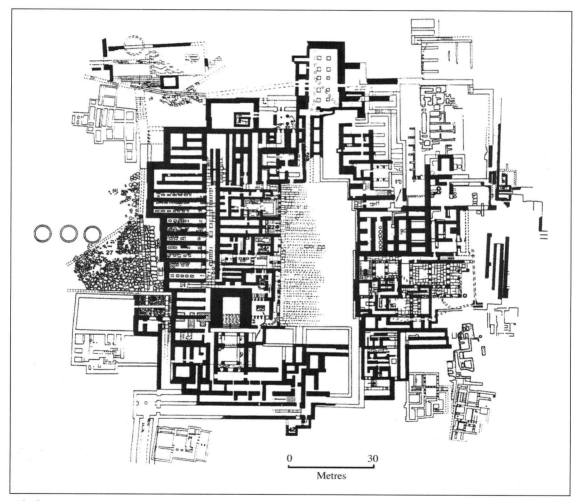

0   30
Metres

Whether or not an Egyptian Labyrinth really existed, there is no doubt that the Greek identification of the probably misnamed Palace of Minos at Knossus in Crete as such had some justification. Its bewildering array of rooms occupied a site some 150m × 150m grouped around a central courtyard. Mud-brick made up much of the construction, possibly with timber framing enabling it to survive earthquakes for 500 years.

The throne room at Knossus, restored in 1930.

Although it is unlikely that Pliny ever saw the Egyptian Labyrinth, he clearly believed it existed, and that Daedalus took it as a model for the Labyrinth he supposedly built in Crete, in which King Minos concealed the Minotaur. In fact it was the Cretan Labyrinth, not the Egyptian one, which existed only in the minds of the Greek writers, who identified it with the palace at Knossos, excavated by Evans between 1900 to 1905.

The Minoans built great palaces at Knossos, Phaistos, Mallia and Zakro and possibly at one or two other sites in the western half of the island; but despite the abundance of stone and timber, the builders used mud-brick extensively in their construction. They faced the walls on the outside with squared stone blocks, frequently gypsum, carefully finished to give a close fit along the front edges only, resulting in a structural weakness which may have contributed to the destruction of these palaces by earthquake around 1700 BC. Rapid reconstruction followed, only for the buildings to be destroyed again around 1450 BC, with the apparent exception of Knossos, which survived until about 1375 BC. It experienced occupation by Greeks from the mainland during its latter years. The exact cause of the last phase of destruction, which, in the case of the mid-fifteenth-century BC disasters, came a generation after the vast volcanic explosion which blew apart the island of Santorini, remains a subject of argument. As foundations and lower walls alone remain of these buildings, the structural features are a matter of intelligent deduction and guesswork. Mostly two-storied, but occasionally three- or four-storied, the upper walls above first-floor level appear to have been constructed of timber framing infilled with mud-brick and rubble, a design evidently intended to resist earthquakes; if so it seems to have been successful as the palaces survived for over 500 years, with only one known major occurrence of destruction by earthquake before the final devastation. Internal columns were of tapered wood with the narrow end downward, the larger diameter at the top providing a substantial support for the heavy timbers, which supported the flat roofs. Although the palaces contained a jumble of rooms, the Minoans showed great ingenuity in utilising stairways and inner courts as light wells. They also built curved gutters beside the open stairways to control and collect the rainwater falling on the building. Facilities within the palace included bathrooms, bathtubs and sanitary systems flushed by a continuous flow of water. The latrines connected to an extensive drainage system through the palace, utilising baked clay or terracotta pipes. Tapering of these pipes gave the water a shooting motion that helped to keep the successive lengths clear of sediment.

When Assurnasirpal II succeeded his father as Assyrian king in 883 BC, and initiated the dominant neo-Assyrian period lasting 270 years, he moved the capital from Assur to Nimrud (ancient Calah or Kalhu), and by so doing set something of a precedent for later Assyrian kings. Sargon II (721–705 BC) moved the capital to a new site at modern Khorsabad (ancient Dur-Sharrukin) and his son Sennacherib (704–681 BC) moved it again to Nineveh, an ancient city dating back to the sixth millennium BC. In one of his many inscriptions Sennacherib wrote:

> I set up great slabs of limestone around its walls [the city walls] and strengthened its structure: over these I filled in the terrace to a height of 170 courses of brickwork [mud-brick], I added to the site of the former palace . . . On this I had them build a palace of ivory, maple, boxwood, mulberry, cedar, cypress, spruce and pistachio the 'The Palace Without a Rival' for my royal abode . . .

In fact, as with all the Assyrian palaces, thick mud-brick walls constituted the main structural feature, with fired brick used for special purposes, such as archways over doors. Cedar beams were greatly favoured for the roof structures. Within the walls of the structures which he commissioned, Sennacherib provided space for libraries and archival material, and for scholars to work at copying Sumerian and Akkadian texts, some of them already over 2,000 years old. Perhaps Sennacherib's example inspired Assurbanipal (668–627 BC) who, during his lengthy period as king, acquired, and housed in his personal library at Nineveh, the most important collection of cuneiform tablets that has ever been found, which came from the length and breadth of his realm. These tablets included masterpieces such as the Epic of Gilgamesh and the Creation Epic. Proud of his intellectual attainments, real or otherwise, Assurbanipal recounts in a famous inscription, his ability as a schoolboy to solve complex mathematical problems and to read abstruse clay tablets written in Sumerian and Akkadian.

Each incoming Assyrian king put in hand large-scale building works, not least in the building of lavish palaces, which, in contrast to Sumerian cities, appear to have taken precedence even over the building of temples. At Kalhu, Assurnasirpal located his city within an area of about 360 hectares, roughly rectangular in shape and bounded on two sides by the Tigris and an irrigation canal, surrounded by fortified mud-brick walls extending to 7.6m in length. His palace, with its attendant temples, was located in a separate walled enclosure, known as the Citadel, at the north-west corner of the site. Raised on a platform 15m high, formed from 120 courses of mud-bricks, it must have been a formidable sight to the general populace living below it. They had their moment, however: on the occasion of its opening, according to a stele found in the ruins of the palace, the king laid on a party for 69,574 people, which lasted for ten days. The palace served not only as the king's residence, but also as an administration centre.

Assurnasirpal II's son Shalmaneser II built a second raised citadel in the south-east corner of the site, comprising a military arsenal with workshops and storage rooms, administrative and residential areas, two courtyards, ceremonial rooms and a throne room with access to a raised dais overlooking a parade ground. As a result of the excavations by Mallowan, and the finding of cuneiform tablets in some of the rooms, more is known about this complex, both its architectural features and its functions, than any other Assyrian site. Finds included large amounts of arms and armour and war booty, particularly ivory. The pattern of having separate walled areas for the citadel and the arsenal became a model for later kings and was repeated at Khorsabad and Nineveh. Architectural forms, construction methods and decorative features also remained fairly standard throughout the neo-Assyrian period. Successive kings built other palaces on the Nimrud site and Assurnasirpal's palace subsequently found various uses as a treasury, granary and as a stopover offering accommodation for caravan traders.

The palace itself consisted of a large number of rooms set out around courtyards, one of which gave access to the most important room, the 47m-long and 10m-wide Throne Room. Past excavations at the site have, not surprisingly, focused more on the palace decorations and adornments than its architectural features. A pair of huge human-headed winged bulls, sculpted from massive stone blocks hauled to the site from quarries near Mosul, guarded each of the entrances to the Throne Room. The provision of five legs allowed these to be viewed as four-legged beasts from both the front and the side. Within the room itself, and also in a number of adjacent rooms, carved and brightly coloured gypsum slabs (orthostats) were set against the lower levels of the thick mud-brick walls, depicting the king vanquishing his enemies in wartime and displaying his prowess at the hunt in peacetime. He is also shown making suitable offerings and libations to the great god Assur. Discoveries at a minor palace near Nimrud of the famous Balawat Gates have revealed that the massive wooden doors of the palaces rotated on pivot stones fixed into the paved flooring and were secured at the top by stone rings. The doors were decorated with bronze bands, featuring miniature reliefs and cuneiform inscriptions, invaluable to archaeologists.

A puzzling feature is the apparent lack of windows in Assyrian architecture, leading to speculation that some form of clerestory lighting may have been adopted, which could have been provided by a roofing system of timber beams and columns. However, it is known that the Assyrians were familiar with brick vaulting, and this form of roofing cannot be discounted.

After sacking Nineveh and bringing the Assyrian Empire to an end, the Babylonians set about turning their capitol into the greatest city the world had ever seen; but it is possible that their most famous building, and the only one included in the Seven Wonders of the Ancient World, may never have existed. In fact it is likely that the Hanging Gardens of Babylon did exist, but the form they took remains conjectural. Herodotus, who certainly visited the city, makes no mention of them. But on the positive side other ancient writers do mention them and even provide descriptions, although differing considerably in detail.

The Hanging Gardens of Babylon are listed as one of the seven wonders of the ancient world, but it is uncertain whether they even existed, or, if they did, what form they took. They may have resembled this reproduction. Diodorus and Strabo both offer descriptions, but Herodotus makes no mention of them. They may have been created by Nebuchadnezzar for his Median Queen, homesick for her mountainous country. (From Hammerton)

Koldewey excavated vaulted substructures that may have supported the gardens. They were reputedly built by Nebuchadnezzar for his wife Amyhia, a Median princess, who missed the lush mountainous terrain of Iran where she had grown up. Although, in part at least, a political marriage (the alliance of Medes and Babylonians having recently defeated the Assyrians), such a romantic story deserves to be true. Nebuchadnezzar may also have felt a compelling challenge to match the magnificent irrigated gardens and orchards which the Assyrian king, Sennacherib, had created in Nineveh and which he destroyed in sacking the city. Koldewey makes the point that the term 'hanging', translated from the Greek κρεμασός (*kremastos*) and from the Latin *pensilis*, did not necessarily convey the same meaning as attributed to them today. *Pensilia* were Roman balconies, which accords well with rising tiers of gardens.

In his description of the gardens Diodorus relies on the writings of Ctesias, which are no longer in existence. He attributes them to a Syrian king who had them constructed to please one of his concubines:

> The park extended four plethra [about 120m] on each side, and since the approach to the garden sloped like a hillside and the several parts of the structure rose from one another tier on tier, the appearance of the whole resembled that of a theatre. When the ascending terraces had been built, there had been constructed beneath them galleries which carried

These vaults may have formed part of the substructure of the Hanging Gardens or some other important structure.

the entire weight of the planted garden and rose little by little one above the other along the approach; and the uppermost gallery, which was 50 cubits [about 25m] high, bore the highest surface of the park, which was made level with the circuit wall of the battlements of the city. Furthermore, the walls, which had been constructed at great expense, were 22 feet [7m] thick, while the passageway between each two walls was 10 feet [3m] wide. The roofs of the galleries were covered over with beams of stone 16 feet [5m] long, inclusive of the overlap, and 4 feet [1.2m] wide. The roof above these beams had first a layer of reeds laid in great quantities of bitumen, over this two courses of baked brick bonded by cement, and, as a third layer, a covering of lead, to the end that the moisture from the soil might not penetrate beneath. On all this again earth had been piled to a depth sufficient for the roots of the largest trees; and the ground, when levelled off, was thickly planted with trees of every kind that, by their great size or any other charm, could give pleasure to the beholder. And since the galleries each projecting beyond another, all received the light, they contained many royal lodgings of every description; and there was one gallery which contained openings leading from the topmost surface and machines for supplying the gardens with water, the machines raising the water in great abundance from the river, although no one outside could see it being done.

It is possible that the stone beams in the account by Diodorus were in fact laid above the arched roofs of the chambers to provide a level surface on which to build up the bases of the garden beds. The vault widths of 3m or slightly greater, found by Koldewey, accords well with the 3m between walls given by Diodorus, but the 30m side lengths of the whole structure found by Koldewey is only one-quarter of the 120m claimed by both Diodorus and Strabo. However, the German archaeologist, having been misled on a number of occasions, warns against taking too literally statements of measurements given by ancient writers. He stresses, instead, the similarities between the ancient descriptions and his own findings of a building with characteristics different from all others found in his excavations. He notes, too, that the importance of the building is attested to by the use of stone, found in his excavations, in its construction. The only other substantial use of stone revealed by his excavations in Babylon was in part of a wall surrounding one of Nebuchadnezzar's palaces, in paving the Processional Way and in the piers of the bridge across the Euphrates.

Koldewey's identification of this structure as possibly the site of the Hanging Gardens rests on his finding, within the area of its remains, a well with three shafts placed close to each other – a square one in the middle flanked by oblong ones on each side. If he is correct in his belief that this is the source of the water essential for the irrigation of the gardens in this arid climate, then Diodorus is incorrect in stating that the water was drawn from the Euphrates. According to Koldewey the water may have been raised by buckets attached to an endless chain passing over a wheel above the well, rotation of the wheel lifting the filled buckets in one outer shaft and lowering the empty ones in the other. The middle shaft could presumably have given access to the reservoir at the base of the shafts. Koldewey observed such contrivances, called *dolabs*, still in use in the neighbourhood. The water may not have been solely for the benefit of the plants. The air entering the chambers through the leaves of the trees would have been cooled by the continuous watering of the vegetation, making it possible for palace officials to work in these chambers through the heat of the summer. Koldewey observed a similar system in use in Turkish government offices in his time.

By the time Darius I came to the throne of Persia in 522 BC the empire was well established and Cambyses had added Egypt to it. The circumstances could hardly have been better for Darius, who proved to be an outstanding ruler, to indulge his passion for building palaces, notably at Susa, which continued to be the centre of Achaemenid government, and at Persepolis. He had vast resources to call on, as evidenced by his own inscription from his palace at Susa:

> This is the palace I built at Susa. Its ornamentation came from afar . . . And that the earth was dug, and rubble was packed, and brick moulded, the Babylonians, they did it. The cedar timber was brought from a mountain in Lebanon; the Assyrians, they brought it to Babylon; from Babylon the Carians and Ionians brought it to Susa . . . The gold came from Sardis and from Bactria . . . the silver and copper from Egypt . . . the ivory from Ethiopia. The stonecutters who wrought the stone, those were Ionians and Sardians and the brickmakers Babylonians. The men who adorned the walls were Medes and Egyptians . . . May Ahuramazda protect me . . .

Little remains of Darius' palace at Susa. Constructed on a low hill, which had been levelled to give a site area of about 4 hectares enclosed within mud-brick walls, the relatively small area led to a congestion of courts and buildings, comprising royal apartments and service rooms. The Apadana, or ceremonial throne room, stood on its own terrace at a higher level. Susa was the focal point of the whole empire. A network of inns and stables served the king's couriers bringing messages from the far points of the empire.

Although Susa remained the main administrative centre for the Persian Empire, Persepolis was its architectural showpiece. Mostly constructed during the reigns of Darius and Xerxes, later kings made further additions until Alexander the Great sacked and burnt it in 331 BC. It occupies a spectacular site, standing on a platform some 450m by 280m, covering an area of about 13 hectares, and made up of limestone blocks; the basic platform, on which no buildings were constructed, rises to a height of up to 13m above the plains and is partly cut into the mountains which dominate the site and from which the city derived its water supply. A fortified mud-brick wall enclosed the site. The main access to the site, situated in the north-west corner, consisted of a double staircase with converging flights of 111 shallow, 7m-wide, steps parallel to the platform walls and leading to a gatehouse called the 'All Lands', fronted by two huge 10m-high carved limestone piers.

The buildings – palaces, audience halls, barracks and treasury – occupied three higher terrace levels. The winged demons and human-headed bulls guarding the gates, as well as the raised terrace and mud-brick walls decorated with panels of polychrome glazed tiles and relief sculptures, echoed earlier Assyrian architecture. However, the reliefs were much more three-dimensional and detailed than Assyrian examples, reflecting a strong Greek influence. In further contrast to Assyrian architecture, the reliefs are repetitious to the point of monotony, and decorate external façades of the artificial terraces and ornamental stairways rather than the internal walls of the buildings. As an officer in the army of Cambyses,

Built mostly during the time of Darius I (522–486 BC) and Xerxes I (486–465), Persepolis was the architectural showpiece and spiritual centre of the Persian Empire.

Darius spent some time in Egypt and, not surprisingly, Egyptian influences are seen in the architecture, in the column and lintel form of construction and absence of vaulting, in interior column arrangements and in the palm and lotus designs at the base of the columns.

Two major halls have been identified, each 67m square in plan. The most imposing of all the buildings on the site was the Hypostyle Hall or Apadana of Darius and Xerxes in which 36 stone columns, over 2m thick and 20m high, supported a roof of cedar, faced with enamelled brick or faience, bronze and ivory. Metal plates capped the ends of the wooden joists to protect them from the weather. The capitals of the columns consisted of the forequarters of bulls, lions or composite animals kneeling back to back. Any exposed wood was lavishly painted. The king sat on a throne beneath a canopy inlaid with plated gold and silver. Further colonnades of 20m-high columns surrounded three sides of the Hall, the whole inviting comparison with the Hypostyle Hall of Ramses II at Karnak, but much more elegant in appearance. A lowered parapet along the length of the Hall, replacing the fortified mud-brick wall, allowed the king and his guests an uninterrupted view over the adjacent plains.

The other large hall, known as the Hall of a Hundred Columns, started by Xerxes and completed by Artaxerxes I, had, as the name indicates, a larger number of columns, but these were shorter and more slender than those of the Apadana. A further sixteen pillars supported a portico at its northern end, and, built into the flanking masonry piers, two bulls guarded the entrance to the hall. The wall separating the hall from the portico had seven stone-framed windows, other walls having niches instead of windows.

Exploitation of the basic arch form allowed Roman emperors, from the first century AD onwards, to indulge their megalomaniac fantasies for vast space-enclosing structures with ever more complex and increasing numbers of intersecting elements. Vast amphitheatres were raised, incorporating dazzling interplays of classic arches, barrel vaults and cross vaults, for the multiple purposes of providing underground passages and chambers where wild beasts, gladiators and Christians could be held, and above ground supporting tiered seating, for some 50,000 people in the case of the Colosseum in Rome, together with corridors for people to circulate and cross passages for entrance, access and egress. Some seventy large amphitheatres were built throughout the Roman Empire, of which the Colosseum was the outstanding example. Started during the reign of the Flavian Emperor Vespasian, and dedicated in AD 80 during the brief reign of his son Titus, many different materials went into its construction, but concrete predominated, with pumice added to it in the vaults to reduce its weight. Much of the deterioration seen in the structure today has resulted from the quarrying of the marble seating and travertine walling and other materials used in its construction. Roman builders displayed admirable adaptability in constructing these great auditoria, many of them, such as those of El Djem in Tunisia and Nîmes in France, being built entirely of stone.

Known as the Flavian Amphitheatre in Roman times, the Colosseum (a name it acquired in the Middle Ages from a nearby huge statue of Nero) was no doubt seen by Vespasian as a monument to himself and the Flavian Dynasty, but his motives for building it probably went beyond this. A commoner who had come up through the army, where he acquired a reputation as a strong disciplinarian, he sought to restore Rome's pride in itself and bring to an end the moral degeneracy which it had suffered from the excesses of Nero, from AD 54 to 68, and the ineptitude of a number of short-lived emperors who had grasped the imperial purple in the 'year of the four emperors' following Nero's death. He saw the building of great structures as a means of rehabilitating Rome to its former greatness, and ordered the draining of an artificial lake that Nero had excavated in the grounds of his Golden House in order to accommodate the great amphitheatre of elliptical shape, having dimensions 189m by 156m in plan. An apparently honest man of great wit, he enjoyed considerable popularity, despite the strong disciplines he imposed and a rather thrifty or even miserly streak, sorely needed after the excesses of previous emperors. When he introduced a tax on urine, used as a tanning agent, his son Titus protested which prompted Vespasian to hold a coin under his nose and asking: 'Does it smell?' His reply: 'Non olet', has become celebrated as signifying that money is money, wherever it comes from.

Titus enjoyed even greater popularity than his father, but died of a fever after only two years as emperor, to be mourned by the whole nation. It was, however, an eventful two years, which witnessed the destruction of Pompeii and Herculaneum in AD 79 from the eruption of Vesuvius; to be followed shortly after by a great plague, then a devastating fire that raged for three days in Rome.

Although the final touches to the Colosseum came after the death of Titus, he dedicated it during his brief time as emperor, turning the occasion into a national festival lasting 100 days. The word 'arena' derives from the sand spread over the floor to soak up the blood: 9,000 beasts died in the arena over the course of the festivities. Gladiators fought in the arena and, as a special spectacle, the arena was flooded, so they could fight over-water

In order to accommodate the Colosseum, Vespasian ordered to be drained a lake constructed by Nero in the grounds of his Golden House. Built mostly of concrete (with marble and other facings), it seated 50,000 people.

battles. It all added to the popularity of Titus. When he caught fever his brother Domitian had him dunked in a tub of snow, ostensibly to reduce his temperature, but more likely to hasten his death. This would have been entirely in character for Domitian, who coveted the Imperial Purple and who, as emperor, included the persecution of Christians amongst his excesses, which matched those of Nero. In AD 96, after a reign of fifteen years, he died at the hands of an assassin.

Fortunately for Rome, and indeed for all those since who value human achievement in building and construction, two outstanding emperors, Trajan and Hadrian, succeeded Domitian. During their reigns and those of their successors Antonius Pius and Marcus Aurelius, for a total of seventy-three years, Rome attained its zenith of power and prosperity. One of the most remarkable pieces of building put in hand by Hadrian was his villa at Tivoli, some 25km east of Rome, the ruined buildings of which now occupy an area some 600m by 300m. Its many buildings, which included, in addition to lavish living quarters, a theatre, various halls, baths, a stadium, temples and libraries, displayed the full range of Roman building know-how: towering walls, colonnaded porticos, arches, vaults and domes.

Two of the greatest vaulted structures – the Baths of Caracalla and the Basilica of Maxentius – were built, respectively, 100 and 200 years after the time of Hadrian. Caracalla, son of the North African born emperor, Septimus Severus, murdered his brother Geta to ensure his sole possession of the Imperial Purple. He ordered the execution of anyone with the remotest connection with Geta. In a deluge of blood, 20,000 people died.

The Colosseum was finished during the brief reign of Titus, who marked its opening with 100 days of festival in which 9,000 wild beasts perished and many gladiators too, some in naval battles.

Spending lavishly, he quickly exhausted the treasury built up by his father and, as a direct result, introduced an act of historical importance, granting citizenship, in AD 212, to the entire provincial population of the Roman Empire. As all citizens had to pay death duties this proved to be, as intended, a source of immense wealth and made possible the construction of the great Baths that bear his name.

The Baths of Caracalla comprised a leisure centre catering for the pleasures and sensual activities of 1,600 people at any time. Bathing pools, 321m square, were built up on a platform 6m high to accommodate, underneath, the furnaces, hot air ducts and hypocausts. Water flowed into the site from the Marcian aqueduct, first into a huge vaulted reservoir, thence by lead piping to the baths and throughout the whole complex. Huge intersecting semi-circular vaults, 32m high, formed the roof of a hall which dominated the whole complex, while a dome similar to that of the Pantheon roofed the structure which housed the hot baths. Rich interior decorations matched the magnitude of the whole project, with no expense spared on coloured marble, stucco facings, and the extensive use of marble, alabaster and granite for columns, carvings and sculptures. In typical Roman manner, and in direct contrast with Greek architecture, the exteriors presented a relatively plain and austere appearance. In recent times the ruins have been used as an amphitheatre for staging operas and concerts.

Begun by Maxentius, but completed by Constantine in AD 313, the Basilica that usually bears the name of the former served a very different purpose from the Baths as the centre of justice, where all civil litigation took place, and where agreements and contracts were signed before a magistrate. It conformed to a fairly standardised plan for basilicae, with a rectangular shape divided longitudinally into a central nave flanked by arched and vaulted aisles, each separated from the nave by supporting pillars. The basic plan became standard for later Romanesque and mediaeval churches and cathedrals. Three huge cross vaults, each spanning 25m, formed the nave and these, together with the use of transverse barrel vaults over the aisles, enabled the internal area of 80m by 51m to be intruded upon by only four columns. It was the largest unencumbered hall built in ancient times. The aisles were partly subdivided into side rooms that received light from windows in the ends of the transverse barrel vaults. All that remains now are the coffered barrel vaults on the north side, but the impressive size of these bears witness to the dimensions of the huge groined vaults over the nave, the lateral thrusts from which were absorbed by these flanking structures.

The palace of the Sassanian king, Chosroes, at Ctesiphon, has good claims to have been the most remarkable pitched brick structure ever built. It pushed the technique (see Postscript) to the limit in achieving a vault of parabolic shape 35m high, 25m wide and 1m thick at the crown, roofing a hall 50m long, without the need for centring in its

Fed by the Marcian aqueduct, the Baths of Caracalla could accommodate the pleasure activities of 1,600 people at any one time. The scanty remains do no justice to a magnificent structure, which had huge semi-circular vaulted roofing and a dome over the hot baths. It was financed by the taxes accruing from making Roman citizens of the entire population of the Empire in AD 212.

193

The Arch of Ctesiphon, built as a palace for the Sassanian king Chosroes around AD 540, is the most striking pitched brick vault ever built. Parabolic in shape, it is 35m high, 25m wide and covered a hall 50m long.

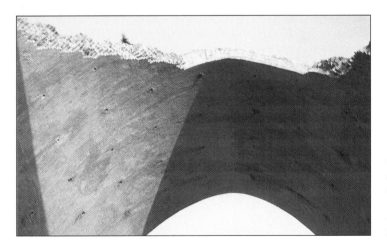

The pitched brick Arch of Ctesiphon is constructed of bricks about 300mm square and 75mm thick, laid back at an angle of 18° from the vertical. It is about 1m thick at the crown. Side walls with a base thickness of 7m take the thrusts of the arch.

construction. Side walls, with a base thickness of 7m, take its lateral thrusts. Having achieved a victory against the Byzantines in 540 BC and capturing Antioch, Chosroes resolved to erect a unique structure in celebration, and he succeeded magnificently. He wisely approached Justinian, the Byzantine emperor, to send experts to him to plan and build the structure. For ceremonial and state occasions the king sat in splendour amid panels moulded and carved in gypsum plaster with figures and stylised plant ornamentation, and decorations that included mosaics depicting his capture of Antioch. His crown, too heavy to wear, was suspended above his head from the vault by golden chains, while his feet rested on carpeting of gold and precious stones with a garden motif.

9

ENTOMBING THE DEAD

Many ancient societies chose to bury their dead, or at least the remains of their more important personages, in cists or chambers under earthen mounds or cairns of stones, often quite large and prominent in the landscape. In some cases the burial chambers were elaborate in their construction, none more so than the Neolithic Newgrange in Ireland, dating from about 3200 BC. Situated on the high point of a long low ridge near the River Boyne, it is one of a cemetery of 'passage graves'. As implied by its designation as a passage grave, the tomb chamber is accessed by a 19m-long, 1m-wide, passage, lined on each side by standing stones or orthostats averaging 1.5m in height, roofed near its entrance by slabs spanning its width, but thereafter by corbelling. The chamber itself is

Newgrange in Ireland, dating from about 3200 BC, is one of a cemetery of passage graves in the valley on the River Boyne. The cruciform central tomb chamber is accessed by a 19m-long passage formed by large standing stones or orthostats.

A gap above the entrance to Newgrange allows sunlight at the winter solstice to shine along the passage and reach the floor of the inner chamber.

cruciform in plan, having recesses on either side and at the back, giving a total length of 5.2m and a width of 6.5m. Its soaring corbelled roof formed from stone slabs, a remarkable achievement for its time, gives the chamber a maximum height of 6m. The passage is oriented south-east–north-west such that a narrow rectangular slit, 20–25mm in width, immediately above the entrance, admits a shaft of light from the rising sun on 21 December, the shortest day of the year, which reaches right through to the tomb chamber for a period of seventeen minutes.

In 1962, when Professor M.J. O'Kelly commenced excavation of Newgrange, the mound (or more correctly, cairn) covering the chamber and access passage had partially collapsed and the form of the mound, now reconstructed, had to be deduced from his findings. It is almost circular and flat-topped, 79–85m across and 11–13m high, and consists mostly of about hand-sized, water-rolled boulders interspersed with layers of *turves*. The external appearance is dominated by an encircling wall over 4m high of white quartz and granite boulders, which retains the outer edge of the mound, and at the foot of which is a continuous kerb of stone slabs about 1.2m in height. A circle of standing megaliths appears to have originally surrounded the mound, probably thirty-five or so in number, but only twelve of these now remain. Many of the kerbstones and passage orthostats are decorated with geometric patterns – such as circles, spirals, lozenges and zig-zags – incised into the stone with a flint point. These decorations bear some similarity

196

to ornamentations in passage graves of a similar character in Portugal, Spain and Brittany, but no direct linkage has been established.

Although common in many parts of Western Europe, passage graves are rare in England, where Neolithic tombs are more commonly elongated stone chambers with recesses on either side containing the bodies, buried under a mound, or long barrow, often ovoid or trapezoidal in plan. As with passage graves the large stones closing the entrance to the chamber could be moved to allow multiple burials, and some held up to at least fifty bodies. West Kennet, occupying a prominent ridgeline position near Avebury, is a particularly impressive example of a long barrow containing a galleried chamber.

The corbelled roof of the Newgrange tomb chamber rises to a height of 6m above floor level.

The tomb chamber is situated at the eastern end of the 100m-long barrow, which is up to 20m wide and 2.5m high. A huge stone blocked the entrance, able to be moved for the multiple burials. The crescent-shaped forecourt forming the entrance was fronted by a row of large sarsen stones, now re-erected. A kerb of sarsens ran around the edge of the mound, most of the material for the mound coming from flanking quarry ditches, 3m deep and 6m wide. Most of the larger sarsen stones, including the main uprights and capstones for the chamber, were brought from Marlborough Downs, 40km to the north, while smaller fill-in stones of oolitic limestone came from a site about 20km to the west.

The tomb itself, which, at 12m, occupies only a short length of the eastern end of the mound, is a parallel-sided gallery with pairs of side chambers on each side 2.3m high with corbelled and capped roofs, and with a fifth chamber at the end. Forty-six bodies, ranging from babies to adults, have been found in the chambers in recent times and others may have been removed in the past, clearly indicating that it was reopened many times, and served many generations over a very long period of time. Pottery, ornaments and food used for ritual purposes during burials and perhaps on other occasions were also found in the chambers.

In Britain the round barrows of the Bronze Age, mostly built between 2000 BC and 1400 BC, of which there are thought to be between 30,000 and 40,000, greatly outnumber the Neolithic long barrows. Unlike the Neolithic communal burials, each round barrow was constructed primarily for one person's remains, occupying a cist below the centre of the barrow, although occasional secondary or satellite burials within a mound have been found, perhaps for lesser or later family members. Many variations have been found in the details of round barrow construction, but essentially, where possible, a circular ditch was dug and the soil thrown inwards to create a mound covering the burial and sometimes also thrown outwards to create a low outer bank. They usually occupy a prominent position, clustered together to form a cemetery of barrows.

Throughout the 3,000 years of pharaonic rule those Egyptians who could afford it concerned themselves to an obsessive degree with their welfare in the after-life. The leading and wealthy citizens took elaborate precautions to ensure their continued existence after death, which they believed depended upon preservation of their earthly body. Their tombs, often constructed of stone or excavated deep into solid rock, were much more elaborate than their homes or even their palaces, which were often constructed for the most part of mud-brick.

An early form of tomb, the *mastaba*, consisted of a burial chamber below ground level, which contained the body, surmounted by a squat superstructure (bench-like, hence the name) of sun-baked mud-brick. This superstructure contained cells for the storage of wine jars, food vessels, hunting implements and other necessities for enjoying the after-life to the full. In some late *mastabas*, of the Fourth Dynasty, stone replaced brick, and this represented a significant development: the interior often consisted of a low-grade local limestone, with an outer facing of fine-quality limestone quarried at Tura.

The Third Dynasty pharaoh, Djoser (2630–2611 BC), had in his court the first multi-talented man in history. Administrator and scribe, sage and writer, engineer/architect and healer of the sick, Imhotep's reputation grew with the telling throughout pharaonic times, leading to his deification by later generations of Egyptians and by the Greeks.

West Kennet long barrow, built around 3000 BC, occupies a prominent ridgeline near Avebury. The tomb chambers occupy only about a 12m length at the eastern end of the 100m-long mound. The massive sarsens forming the tomb chambers and blocking stones were brought to the site from the source 4km to the north. (Jason Hawkes)

Commissioned by Djoser to build him a tomb at Saqqara fit for a god–king, Imhotep first constructed, out of limestone blocks, a *mastaba* 63m square and 8m high, each face of which he oriented towards one of the four cardinal points. Whether Djoser expressed himself dissatisfied with this structure or whether Imhotep, in part at least to satisfy his own ambitions, persuaded him that only something more imposing would befit a god-king is unknown; but he went on to extend the *mastaba*, first into a four-stepped pyramid and then, finally, into a six-stepped pyramid having base measurements of 125m by 110m, and a height of 62m. Technologically a great step forward in masonry design and construction techniques, it ushered in a period of frenetic stone pyramid construction.

In order to create a stable structure, Imhotep built up a central core or nucleus of stone blocks with an inward slope of 74° to the horizontal. Successive buttress walls of stone blocks were added, also inclined at 74°, from the core outwards, each finishing at a lower level than the previous one. A pyramid shape resulted, with a stepped exterior having a slope of 51.9° to the horizontal. Although this was, on the face of it, a logical solution, the planar interfaces between the buttress walls introduced a weakness into the structure.

The next pyramid after the Step Pyramid to be completed, or certainly very nearly so, was sited at Meidum, 50km to the south of Saqqara: probably intended for Huni (*c.* 2599–2575 BC), the last king of the Third Dynasty, about whom little is known. It is now attributed to Snofru (2575–2551 BC), the first pharaoh of the Fourth Dynasty, thought to be the son or son-in-law of Huni. The buttress wall design of this pyramid followed closely that of its Saqqaran predecessor, but a decision, made late in the construction, to provide the pyramid with smooth, rather than stepped, faces proved disastrous. The builders attempted to accomplish this by cladding the exterior with a coating of small local limestone blocks with an outer skin of fine Tura limestone. This, together with the inherent planar weaknesses within the structure, caused the pyramid to collapse. The structure, as seen today, consists of a stark stone column, comprising the core and inner buttress walls, projecting above a mound of sand, rubble and stone chippings, bearing eloquent testimony to the cataclysmic nature of the collapse.

The next pyramid, the South Dahshur or Bent Pyramid, was built on a relatively weak foundation stratum, which caused it to deform during construction. Mindful of the Meidum collapse, the builders finished the pyramid off at a flatter slope. Like Meidum, it could not be contemplated as a suitable repository for the body of Pharaoh Snofru. A solution was needed – and quickly. The pharaoh must have been ageing by now. As civil engineers have throughout history, the builders learnt their lesson from their failures and changed the design of the structure, adopting the flatter slope of 43.5° used to finish off its neighbour and, instead of buttress walls, adopted a coursed form of construction for the North Dahshur Pyramid, building the pyramid up tier by tier, each tier consisting of a single layer of blocks uniform in width across the full width of the structure. It has remained stable to this day.

Pyramid building reached its apogee with the construction on the Giza plateau, on the outskirts of modern Cairo, of the Great Pyramid for Pharaoh Khufu, Snofru's son and successor, who ensured his position by marrying his sister Merytyetes, but about whom little is known otherwise. Herodotus, listening to the priests of his own time in the fifth century BC, presents both Khufu and his son Khafre (mistakenly said by Herodotus to be

The Great Pyramid at Giza. With confidence restored after the successful completion of the North Dahshur Pyramid, Khufu's builders opted for layered construction rather than buttress walls and returned to the favoured steeper slope angle of about 52°. The pyramid has base dimensions of 230m, a height of 147m and its 2.6 million cubic metres volume is made up of some 2.3 million limestone blocks weighing on average 2.5 tonnes each. It was built in about twenty years.

Khufu's brother) as oppressive tyrants. The priests may have based this belief simply on the fact that the two largest pyramids had been built for these two pharaohs, demanding enormous labour inputs, but in fact the total volume of pyramid construction for Snofru exceeded the Great Pyramid by 60 per cent, despite which Snofru enjoyed revered status throughout pharaonic times.

Herodotus gives what is generally a very believable description of how the Great Pyramid was built:

To build the pyramid itself took twenty years; it is square at the base, its height (800 feet) equal to the length of each side; it is of polished stone blocks beautifully fitted, none of the blocks being less than 30 feet long. The method employed was to build it in steps, or, as some call them, tiers or terraces. When the base was complete, the blocks for the first tier above it were lifted from ground level by contrivances made of short timbers; on this first tier there was another, which raised the blocks a stage higher, then another which raised them higher still. Each tier, or storey, had its set of levers, or it may be that they used the same here. The finishing-off of the pyramid was begun at the top and continued downwards, ending with the lowest parts nearest the ground. An inscription is cut upon it in Egyptian characters recording the amount spent on radishes, onions and leeks for the labourers, and I remember distinctly that the interpreter who read me the inscription said the sum was 1,600 talents of silver. If this is true, how much must have been spent in addition on bread and clothing for the labourers during all those years the building was going on – not to mention the time it took (not a little, I should think) to quarry and haul the stone, and to construct the underground chamber?

Although its 230m-square base is only slightly greater than the 219m of the North Dahshur Pyramid, the Great Pyramid with its slope of 52° exceeds by 50 per cent the volume of its immediate predecessor. Its 2.6 million m³ volume (including a small knoll on which it is built) is thought to consist of around 2.3 million limestone blocks, averaging about 2.5 tons each, but generally ranging from about 1 tonne up to 7.5 tonnes or greater. Measurements made by Petrie showed the Great Pyramid to consist of 203 courses, fluctuating in thickness through the height of the structure, but generally decreasing from nearly 1.5m thick at the base to 0.5m approaching the top. The fluctuating thickness probably reflects conditions in the quarries.

The evolution which took place in pyramid design means that no two stone pyramids are the same. They all differ also in the locations and structures of the tomb chambers, and even the numbers of tomb chambers. Some were excavated into the rock strata underlying the pyramid and others located within the pyramid itself. The Great Pyramid has an unfinished chamber below the pyramid and two chambers, the wrongly named Queen's Chamber and the King's Chamber, within the structure, the latter almost at mid-height. The King's Chamber also features five weight-relieving hollow compartments immediately above it. The unique Grand Gallery, one of the great architectural achievements of the ancient world, accesses the chamber itself.

Most Egyptologists agree with Herodotus that it took twenty years, or thereabouts, to construct the Great Pyramid, which means that throughout the construction period, on average, one block had to be placed every two minutes. This probably means blocks had to be placed at a rate of more like one every minute in the lower, more easily accessed, layers. These logistics applied equally to the quarrying, transportation to site, raising of the stones up to the working level and manoeuvring them into place.

Most of the stones for the Great Pyramid came from the Giza plateau itself, which consists of a relatively soft limestone, with the better-quality limestone used for the outer casing coming from underground quarries at Tura, on the other (eastern) side of the Nile, thus requiring the blocks to be transported many kilometres over land as well as across the Nile. Sledge usage to transport and raise the stones has been proposed or favoured by most Egyptologists writing on the subject. The Egyptians certainly used sledges but probably for the most part for one-off operations in which time and economic usage of labour were not factors. On the other hand, the need to place, on average, one block every two minutes in constructing the Great Pyramid (as well as its immediate predecessor and successor) made these two factors of the greatest importance, and, at the same time, make it unlikely that highly inefficient sledges were used in transporting or raising the stones. It is much more efficient to roll a heavy object than it is to drag it, and it is a relatively easy matter to convert an object of reasonably regular shape, such as the pyramid blocks, into a

This simple cradle-like device, in the form of a quarter circle, may have been the high technology of the Pyramid Age. Attaching these cradles to heavy stone blocks would have allowed them to be rolled, a far more efficient way than using sledges to transport and raise them.

form capable of being rolled. The contrivances made of short pieces of timber described by Herodotus may have been used for this purpose. The obvious artefact fitting this description is a cradle-like device, consisting of side pieces with curved undersides connected by an arc of dowels, many models of which Petrie found in the foundation deposits of Queen Hatshepsut's mortuary temple. The curved undersides have the geometry of a quarter-circle. Fitting four of the quarter-circles around a block created a circular runner and fitting two of these runners to a block allowed it to be rolled, thus creating a highly efficient means of transporting and raising it. The advantages over sledge usage include:

1. The tracks required between quarries and ramps and on the ramps themselves needed only to be firm and even, with no other special preparation. Unlike the sledge the rolling technique needs a good natural frictional surface, not a lubricated one.

2. On a level surface a 2.5-ton stone block can be propelled by two or three men at a fast walking pace. Under field conditions of pyramid construction, with so many blocks in transit at any one time, it is unlikely that, even with special track preparation and

Three men rolling a 2.5-tonne concrete block fitted with cradle runners on a level, lightly compacted, gravel surface at a fast walking pace. The ropes are required only for negotiating slopes. About twenty men would have been required to pull the same block on a sledge over a level surface, which would also have needed special preparation to reduce friction, possibly by inserting timber cross-pieces into the road surface. As moving blocks from the quarries to the site was equally important as raising them, the ease of rolling, and the requirement of only minimum road surface preparation, would have given this method considerable advantage over the use of sledges.

friction reduction, fewer than sixteen to twenty men could have hauled the same stone, and then only slowly and laboriously. The advantage of rolling on a level surface becomes even greater with heavier stones.

3. The steepest slope which can feasibly be contemplated using a sledge is 1 in 10, but it is difficult, if not impossible, to design a ramp system with such flat slopes. Slopes as steep as 1 in 4 can be negotiated by rolling the stones, thus making possible a practicable ramp system. The number of men required to roll a stone up a 1 in 4 slope is about half the number required to drag the same stone on a sledge up a much flatter 1 in 10 slope.

4. Hauling a sledge up a slope with a low friction contact is hazardous as there is no simple way to stop the sledge careering out of control down the slope if the ropes break or slip in the hands of the haulers. A rolling stone can easily be stopped and held on a slope by chocking it.

5. It is most likely that the ramps used to raise the stones spiralled around the pyramid in some manner, which would have effectively precluded the use of sledges, as there is no obvious way they could have been manoeuvred around the corners. Rolling stones can easily be manoeuvred onto a central block and swivelled around through 90°.

Many ramp systems have been proposed, most of them impracticable or impossible to construct. Ramps meeting a pyramid face at right angles would have led to ramp systems of excessive length and great volume, which could have been avoided by the use of ramps somehow attached to one or more faces of the pyramid. While spiral ramps of this type have been proposed, it would have been extremely difficult to have created such a system with ramp slopes as flat as 1 in 10 needed for sledges. However, the adoption of steeper slopes of 1 in 4, negotiable by rolling stones, would have made such a system not only eminently possible, but also economical in the use of materials. By placing desert sand and rubble, possibly with the inclusion of quarry chippings, at its natural angle of repose against the four sides of the pyramid, a bank of material could have been formed on top of which ramps of any width or slope could have been created, climbing up against the pyramid faces. Ramp slopes of 1 in 4 would have allowed this rubble bank to reach to

Rolling a 2.5-tonne concrete block up a 1 in 4 slope. Ten men were able to move the block a short distance and fourteen men could haul it up the full 15m ramp length, although up to twenty men were used. About fifty or sixty men would have been required to haul the same sledge-mounted block up this slope.

one-third the final height of the pyramid, by which time two-thirds of the volume of the structure would have been completed with ramps of sufficient width to allow simultaneous raising of many stones. Above this, spiralling stone ramps could have been used, resting on the completed steps of the structure (which would have been trimmed off, together with the removal of the ramps as the final operation in construction of the pyramid). One of the rubble ramps, and a portion of stone ramp, could have been dedicated to raising the massive granite blocks used for the King's Chamber. The sand and rubble mound around the base of the failed Meidum Pyramid may well be the remains of the lower part of such a ramp.

Excavation of the Royal Graves of Ur by Woolley revealed the use of both domed and vaulted underground structures to house the remains of the dead. These date from about the middle of the third millennium BC, and thus correspond closely in age with the great stone pyramids in Egypt. A typical royal tomb consisted of a domed tomb chamber, of limestone construction, some 2–3m in internal diameter, buried under a fill of clay and broken brick at the base of a 15m-deep shaft, above which, on a levelled surface of the fill, mud-brick walls had been built to enclose a ritual area about 5m in width embodying a vault of radial arch, mud-brick construction, about 2.5m internal diameter. The bodies of kings and queens and other exalted personages occupied the domed and vaulted tomb chambers. Further fill, in which individual graves of lesser personages were found, covered the walled enclosure and vault, up to modern ground level.

Rich offerings in gold, silver and bronze accompanied the bodies of the kings and queens and other high-ranking individuals to their final resting place, together with food contained in clay vessels. More gruesomely, the ritual of the interment included the burial alive of animals and humans. Kings and queens were buried with a small number of specially favoured attendants inside the tomb chamber, while in courtyards outside, buried under the fill, excavators have found ladies of the court in all their finery, with head-dresses of gold, carnelian and lapis lazuli, officers with golden weapons befitting their senior ranking, musicians with harps, lyres or silver flutes, soldiers of the guard with their weapons, and even the royal chariot complete with its span of asses or oxen, accompanied by their grooms. Although they no doubt considered it a privilege to be buried in this way with their god–king or his consort, the attendants nevertheless appear to have been drugged with narcotics once they had assumed their appropriate stations in the pit or on the ramp leading down to it. Their burial did not end the ceremony. As the filling reached various levels it was trodden flat to form a firm floor on which a funeral feast was held.

Corbelled construction seems to have found particular favour for the building of underground tombs in the ancient world, reaching an apogee in the great tholos tombs of Mycenae. Contemporary with these were the tombs of finely dressed and beautifully fitted limestone blocks which wealthy merchants of the Phoenician harbour city of Ugarit had constructed beneath their houses. Stone stairs led down to the entrance, and a stone-lined opening in the top of the vault enabled the family to pour in libations to appease the spirits of their ancestors. Vases of delicate eggshell ware from Knossos excavated at Ugarit show it to have been in touch with that Minoan city as early as 1900 BC. Later, around 1400 BC, it came strongly under the influence of Mycenae, and many of the objects found deposited with the dead were Mycenaean.

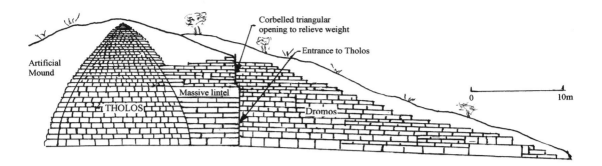

Corbelled tomb construction reached its apogee in this tholos tomb at Mycenae, dating from around 1400 BC, and known variously as the Treasury of Atreus and the Tomb of Agamemnon.

It is possible that the great beehive tombs of Mycenae had their origins in the round tholos tombs in Crete, although the latter, unlike the former, stand in the open plain, and, from the relatively small stones employed, it is questionable if corbelled construction had been used for the roofs. There are, however, remains of very primitive tholos tombs in both Crete and the Cyclades that, as they used rather larger stone sizes, could well have had corbelled stone roofs.

Prior to the construction of the Pantheon in Rome, the Mycenaean tholos tomb called the Treasury of Atreus by the Greeks and more recently dubbed the Tomb of Agamemnon (although neither name has any supporting evidence) had the distinction, for nearly 1,500 years, of being the largest man-made domed structure in the world. It comprised one of a number of similar structures set in the hillsides close to the walled fortress city of Mycenae and was constructed between 1500 BC and 1300 BC, a period coincident with Mycenae holding reign as the leading maritime power, its ships raiding and trading throughout the Mediterranean.

Preparations for the construction of a tholos tomb consisted of constructing a long horizontal entrance passage or *dromos* giving access to the base of a deep vertical shaft sunk into the hillside. Rings of horizontal stones were then placed to build up the structure, each ring having a diameter slightly smaller than the ring below. The space between the structure and the shaft was progressively backfilled with earth. The rough flat stones commonly used in early examples had given way, by the middle of the fifteenth century BC, to carefully dressed stone blocks producing a smooth interior. The pointed, beehive-shaped chamber evolved naturally, as a rounded top would have been difficult to achieve with corbelled construction. In some cases faulty estimates of the required depth of shaft resulted in the top of the chamber wall projecting above the natural hillside and a covering of earth or a stone parapet had to be provided to protect it. At the entrance from the *dromos* into the Treasury of Atreus tholos, a lined façade 10m high houses a 5.4m-high door surmounted by a heavy lintel weighing over 100 tonnes, the load on which was

Entrance to the Treasury of Atreus tholos tomb at Mycenae.

relieved by a corbelled triangular space above it similar to that above the Lion Gate entrance to Mycenae itself. Setting the stone lintel in place, no doubt by manoeuvring it down the hillside, constituted a major engineering feat.

The finished chamber had an internal diameter of nearly 15m and a similar height made up of thirty-three courses of stone. Nail holes, still visible in the face, held bronze rosettes and other ornaments. Unfortunately, the prominence of these tombs with their inviting entrance passages, lined with dressed stone walls, inevitably attracted the attention of tomb robbers, so nothing is known of their occupants, their contents or even the rituals which may have taken place within the interior. The demise of Mycenae around 1200 BC,

Corbelled tomb chamber of the
Treasury of Atreus.

however, did not spell the end of tholos tombs in the Mediterranean. The Etruscans, perhaps influenced by Mycenaean examples, created some impressive underground, corbelled tombs in the seventh century BC.

The most massive tomb structures outside the pyramids were the huge earth mounds built by the Lydians to protect the mortal remains of their kings. A very rich society, boasting none other than Croesus himself as their last king, they could afford such constructions. They clearly impressed Herodotus, who would certainly have seen these enormous mounds, situated on the Ionian coast close to Samos, where he spent some time as a young man. He described, in some detail, the tomb of Alyattes (617–560 BC), the father of Croesus:

> The country (Lydia), unlike some others, has few natural features of much consequence for a historian to describe, except the gold dust which is washed down from Tmolus; it can show, however, the greatest work of human hands in the world, apart from the

This huge mound near Sardis in modern Turkey mentioned by Herodotus contains the tomb of a Lydian king, said to be Alyattes. It has the same height, but a larger base diameter, as Silbury Hill, the largest ancient artificial mound in Europe.

Egyptian and Babylonian: I mean the tomb of Croesus' father Alyattes. The base of the monument is built of huge stone blocks: the rest of it is a mound of earth. It was raised by the joint labour of tradesmen, craftsmen and prostitutes, and on the top of it there survived until my own day five stone pillars with inscriptions cut in them to show the amount of work done by each class. Calculations show that the prostitutes' share was the largest. Working class girls in Lydia prostitute themselves without exception to collect money for their dowries, and continue the practice until they marry. They choose their own husbands. The circumference of the tomb is nearly three-quarters of a mile, and its breadth about 400 yards. Near it is a large lake – the lake of Gyges – said by the Lydians never to be dry.

Although Herodotus exaggerates the size of the mound, excavations have confirmed, in general terms, his description. In fact the mound has a height of 40m and a diameter of 198m, making it marginally bigger than Silbury Hill, the largest ancient artificial mound in Europe, which is identical in height, but has a base diameter about three-quarters that of the Lydian mound. Unlike its counterpart in modern Turkey, however, there is no evidence that Silbury Hill was constructed to cover a tomb.

A Note on Building Materials, Structural Forms and Classical Column Orders

One of the ways in which human technological resourcefulness has manifested itself most obviously has been in the exploitation of naturally occurring materials for building purposes. Animal, vegetable and mineral have all been pressed into service. Even today primitive structures are still constructed using the materials readily to hand. Indeed, the great variety of these structures gives some indication of the range and types of shelters built in ancient times. Most of these had a very short life span, so that no evidence of their existence survives today. The simplest shelters would have consisted of saplings or bamboo canes with a cladding, consisting of animal skins, tree bark, palm leaves, thatch, woven reeds or grasses, or perhaps woven animal hair, to keep out the elements. Where more permanent or substantial structures were required, heavier timber frames would have been used. Wattle and daub structures have an ancient heritage and the technique has survived despite Vitruvius' fervent cry that he wished it had never been invented, citing its proneness to fire and the development of cracking in the walls.

For ancient civilisations in Egypt and Mesopotamia, growing up close to major rivers, the materials that were most readily available, and suitable for using as building material, comprised mainly mud and reeds. In the marshes of southern Iraq remarkable structures using an ancient building technique can still be seen, consisting of parabolic arched ribs formed from tightly bound bundles of reeds, with the ends of each 6m-long rib firmly embedded to a depth of 0.75m into the ground. The parabolic vault formed in this way is covered with woven reed matting fixed to horizontals, formed from reed bundles, and secured to the ribs. Most reed-built structures were much simpler and more primitive than this. Reeds plastered with mud to give a conical or domed shape were common and are still built in Africa and other parts of the world today. Reeds and palm leaf strips could also be formed into matting to reinforce mud-brick structures, as seen today in the Aqar Guf ziggurat near Baghdad, where the builders laid mats over each sixth layer of mud-brick to enhance its lateral strength and possibly to aid in draining excess moisture from the structure.

Tightly bound reeds are used for the structure of this meeting house in southern Iraq.

Early settlements with nearby access to forested areas made use of timber as a building material, either on its own or as a framework to support the roof and wall cladding of thatch, woven reeds or leaf matting, mud or other available material suitable as a barrier against the elements. In hot arid countries such as Egypt and Mesopotamia, the light timbers available locally, such as date palm, fig, acacia, sycamore and tamarisk, although not suitable for heavier building and joinery work, found some application, for example, in roofing minor structures where spans did not exceed 2–3m. Timber for major construction works had to be imported: cedar from Lebanon was most in demand for this purpose. Surviving fragments of the Palermo stone, carved in Fifth Dynasty Egypt, mention the importation of forty shiploads of cedar from Lebanon in one year during the reign of Snofru, to whom three major stone pyramids are attributed. Much of this would have been intended for shipbuilding and for constructing the cradle-like devices (or, as some believe, sledges) for transporting and raising the pyramid stones. Cedar timbers were also used by the builders of the Bent Pyramid in a desperate bid to stop the distortions in the upper tomb chamber arising from settlement of the whole structure. The Persians used large amounts of Lebanon cedar in their palaces, particularly for roofing beams and panelling, lavishly coating the exposed wood with paint, which, not being oil based, would have needed frequent application. In spacing the columns of the Apadana at Persepolis (almost 9m from centre to centre), the cedar beams of the roof spans between them, allowing for capitals, would have had unsupported lengths of about 6m, which the builders would have known from experience could not be safely exceeded with timber beams used in this way.

The Minoans, Greeks and Romans had ready access to a variety of timbers and the Minoans made good use of this gift of nature by adopting, for their palaces, timber-framed walls infilled with mud-brick or rubble to give a flexible, earthquake-resistant, design. The Roman writer Vitruvius in his treatise on the *Ten Books of Architecture* gives a detailed account of the relative qualities and uses of various timbers; these depended, according to him, on the varying amounts of the four basic elements – air, fire, earth and water – present in each of them. On fir:

> it contains a great deal of air and fire with very little moisture and the earthy . . . it does not easily bend under load and keeps its straightness when used in the framework. But it contains so much heat that it generates and encourages decay . . . and it also kindles fire quickly because of the air in its body . . .

and on oak:

> Oak, on the other hand, having enough and to spare of the earthy among its elements, and containing but little moisture, air and fire, lasts for an unlimited period when buried in underground structure.

While the concept of the four elements may seem absurd in the light of modern knowledge, it gave the Greeks and Romans a basis on which to comprehend the relative properties of the materials they worked with, which they knew well enough from experience. Interestingly, Vitruvius dismisses alder as a building material, but points out, as is well known today from finding surviving Roman examples, that underground piles remain imperishable forever if permanently surrounded by moisture (i.e. groundwater).

There is evidence from the writings of Vitruvius that the Romans, by the first century BC, used the triangulated truss to overcome the span limitations of the simple timber beam. In his description the two main rafters are connected at the ridgeline and form the sloping (compression) members of each truss, while a beam spanning the columns or walls, and secured to the rafters at each end, forms the tie beam and completes the triangulation. His description lacks clarity, but for longer spans he seems to propose a system of cross-beams and vertical struts, connecting the tie beam to the rafters, deeming these to be not necessary for shorter spans. In fact, vertical struts play little part in carrying the roof load, but can help the truss maintain its shape. It is not necessary for the tie beam to be a single member, and by using shorter lengths with suitable jointing much greater spans can be achieved than with simple timber beams.

The only known pictorial representation of a Roman timber truss is in a document held in the Palace of the Canons in the Vatican, which shows in section the first basilica of St Peter built around 330 BC. The truss spans the full 24m width of the nave and, in addition to the basic triangulation of rafters and tie beam, has a vertical king post and an intermediate tie beam, known as a straining beam.

The use of stone in ancient building was often confined to tombs and temples, while more secular buildings, including royal palaces in some cases, were commonly built of timber and/or sun-dried mud. Because of its perishable nature and, often, destruction by fire, timber remains are scarce in the archaeological record, but many sun-dried mud buildings have survived sufficiently well for the nature of the structures, their usage and even their building techniques to be deduced.

Early builders employed various techniques, with differing degrees of sophistication, to create mud or earthen structures, the simplest of which consisted of raising the walls of a building progressively in layers using pats of mud, usually mixed with straw, reeds or grass to improve its strength and to reduce shrinkage during drying. Placement of each layer, about a quarter of a metre thick, started at one corner and progressed around the building, the new pats being trodden down onto the partly dried and firm previous layer. In some cases no formwork was used, and the walls were simply trimmed of excess material using a spade-like tool. The technique could, however, be adapted for use with formwork, thus eliminating waste and the need for trimming; with solid formwork relatively dry mixes could be used, thus reducing the amount of shrinkage on drying.

Evidence from sites such as Jericho and Catal Hüyük shows that the use of mud-bricks dates back to at least as early as the eighth millennium BC in southern Anatolia and the Levant. By the fourth millennium BC mud-brick structures abounded throughout the Middle East, Greece and Egypt. In addition to utilising the most plentiful material available, the poor heat-conducting properties of mud-bricks made them particularly suitable for hot arid climates. The temperature inside a well-constructed mud-brick house probably rarely exceeded 25 °C even when the outside temperature reached 35 °C.

Mud-bricks were made from soil, preferably with some clay content, by bringing it to a plastic consistency with the addition of water (if too dry) or partially drying it (if too wet). Chopped straw, grass or other organic fibrous material was added to impart strength and reduce shrinkage on drying, and the bricks shaped either by hand moulding or by pressing the plastic mud into moulds. In the former method various shapes could be achieved according to tradition, pear and bun shapes, as well as rectangular shapes, being quite common. Shapes other than rectangular gave the possibility of better interlocking of the bricks and better integration with the mud mortar. Pear-shaped blocks, still used by the Hausa in northern Nigeria, were placed with their wide bases downwards, three or four across the width of the wall, with the bases of the next layer interlocking between their narrow tops. In order to reduce shrinkage and cracking of the wall, the blocks were usually baked in the sun before use.

Mud-bricks formed in a mould were usually rectangular in shape, although in Mesopotamia in the period 2800 BC to 2300 BC a distinctive plano-convex shape became fashionable, with a planar base and planar sides, but with the top rounded off, by hand, proud of the mould. Another distinctive technique consisted of laying the bricks in a herringbone pattern with horizontal courses. Brick sizes varied greatly, with lengths commonly around 300mm, but ranging from 200 to 400mm, with width often about one-half the length and thickness ranging from about 50 to 100 mm. Where moisture-sensitive sun-baked mud-bricks were used in large public structures such as ziggurats, kiln-baked bricks set in bitumen mortar were commonly used for the exposed faces to resist the weather. Ancient builders developed great skills in the building of mud-brick structures and at sites such as Catal Hüyük multi-storey structures evolved, not dissimilar to the Indian pueblo villages found today in the arid south-west of the United States.

Rising damp in the base of the walls and the provision of an adequate roof posed problems for these early builders using mud-brick. Rapid deterioration of the base of a mud-brick wall, from rising damp or the splashing of rainwater, was combated in some cases by constructing the bases and foundations in stone or burnt brick. Roofing presented an even greater problem as flat mud slabs, even with fibrous reinforcing, had limited flexural strength; but nevertheless flat roofs of limited spans were constructed in this way in the absence of suitable timber. This inevitably resulted in small and congested enclosed spaces. The problem could be overcome by keeping the mud-brick in compression, and the imagination and great subtlety shown by builders in creating domed, beehive and conical shaped buildings, to utilise the strength of the material to the greatest effect, demands our admiration today.

Where timber flourished, roofing presented little problem. A sloping roof of close-set timbering could be provided, covered by a layer of thatch, mud or tiling. If timber was available but scarce and precious, spaced timbers could be used expeditiously to reinforce roofs composed of a matting of plant fibres or reinforced mud slabs, or a combination of these. A common sight in parts of Africa today is the circular mud house with a conical timber-framed thatched roof.

Factories mass producing mud-bricks must have been a common site in many parts of the ancient world, as today in this example near Cairo.

In Egypt the availability of limestone led to its widespread use in the construction of important structures such as the major pyramids, temples and palaces; but mud-brick was commonly used for secondary structures and enclosure walls associated with these large edifices. Most Egyptian people also lived in simple mud-brick houses. Not infrequently mud-brick was used for the major structures themselves, not least in the construction of many important pyramids over a period of some hundreds of years following the end of stone pyramid building.

The appeal of mud-brick lay in its availability and its ability to be moulded, and hence there were no difficulties with quarrying and dressing, or with transportation or raising large blocks, or speed of construction. Having ditched Amon and other Egyptian gods in favour of a single deity, the Aten, represented as a sun disc with radiating rays, and having changed his name from Amenhotep IV, Pharaoh Akhnaten needed a new capital city in a hurry, divorced from and free of the priesthood of Thebes. He chose a site midway between Thebes and Memphis and, in the interests of speed, had his vast palace, other public buildings and private dwellings built largely or wholly of mud-brick. The remains of one of the administrative buildings contained correspondence between the king and rulers in Asia Minor, consisting of a large number of clay tablets known as the 'Armana Letters'. Much of the correspondence came from Egypt's provinces in Syria, desperately asking for help against the Hittites. But Akhnaten was too busy worshipping the Aten to attend to such worldly affairs. Occupation of the mud-brick city of Akhetaten lasted only for the last eleven years of Akhnaten's life with the capital returning to Thebes and to Amon shortly after his death.

Fired bricks and tiles can be made from a wide variety of clayey materials, but should preferably contain up to 30 per cent silt, or preferably sand, to reduce shrinkage when the bricks are burnt. The alluvial clays deposited by the rivers of Mesopotamia and Persia proved to be excellent for brick making, containing large amounts of silt as well as some wind-blown sand and limestone dust, the last giving the final product a cream or beige colour. Some of these clays also contained gypsum, which could cause efflorescence on the face of brickwork or cracking of the bricks, but as it occurred in the form of crystalline nodules the ancient brick makers learnt how to avoid it. Davey reports that tests he performed on surface clays in Southern Persia fired at 950 °C produced sound bricks, whereas clay taken from two feet down, and containing more gypsum, produced poor-quality, fissured bricks.

Ancient civilisations used far more sun-dried bricks than fired bricks despite the shortcomings of the former, which included deterioration under adverse weather conditions, particularly alternate wetting and drying, and the fact that they did not readily lend themselves to lavish decoration. Good-quality fired bricks did not have these drawbacks and the technique of firing bricks was known even to the early Mesopotamian civilisations; but a lack of suitable fuel made large-scale production of these prohibitive. Consequently they found their main use as facing bricks, either set in bitumen to protect the mud-brick structure from the weather, as in the ziggurats or defensive walls, or to enhance the decorative possibilities, as in the Ishtar Gate, which featured specially glazed facing bricks to give the famous striking animal reliefs. Nebuchadnezzar was unusually profligate in the use of fired bricks in his great rebuilding programme for Babylon, partly as a response to rapid deterioration of sun-dried brick structures built during the reign of his father Nabopolassar.

Fired bricks also found use in otherwise mud-brick walls in locations particularly susceptible to wetting, such as along the base or the top of the walls. Vitruvius describes such usage for dwellings:

> they (the walls) should be constructed as follows in order to be perfect and durable. On the top of the wall lay a structure of burnt brick, about a foot and a half in height, under the tiles and projecting like a coping. Thus the defects in these [mud-brick] walls can be avoided. For when the tiles on the roof are broken or thrown down by the wind so that rainwater can seep through, this burnt brick coating will prevent the crude brick [mud-brick] from being damaged, and the cornice-like projection will throw off the drops beyond the vertical face, and thus the walls, though of crude brick structure, will be preserved intact.

The Romans probably learnt the basic techniques of baked brick manufacture from the Etruscans or Greeks, but did not capitalise on this knowledge until late Republican times. Vitruvius in his time clearly had little confidence in the quality of the bricks being produced:

> With regard to burnt brick, nobody can tell offhand whether it is of the best or unfit to use in a wall, because its strength can be tested only after it has been used on a roof and exposed to bad weather and time – then, if it is good, it is accepted. If not made of good clay or if not baked sufficiently, it shows itself defective there when exposed to frost and rime. Bricks that will not stand exposure on roofs can never be strong enough to carry its load in a wall. Hence the strongest burnt brick walls are those that are constructed out of old roofing tiles.

The principal difference between Roman fired bricks and tiles arose from the baking time; baking the tiles for longer made them more waterproof, while light baking of the bricks left them with a degree of porosity, enabling them to absorb moisture from the mortar and ensure a firm bond. Roman brick makers produced their baked bricks in three main sizes: 200mm, 440mm and 600mm square and generally 35–45mm in thickness. It was not customary to construct the walls of buildings entirely of brick; most walls had a lime-concrete interior with a facing of embedded bricks. Triangular-shaped bricks were often used for this purpose, cut from square burnt bricks, and embedded with the apex inward, occasionally reversing the embedment to give an attractive textured wall. The process of embedding bricks into lime-concrete in this way also meant that good use could be made of old broken brick and tile rubble. Every metre or so in height a row of bricks spanned the full length of the wall for the purpose of taking the weight of the wall through the brickwork during the lengthy period it took for the lime-concrete to harden.

The technique of burning limestone to produce quicklime, which could be slaked with water to create lime suitable, when combined with sand or rubble, for making mortar or concrete, was known throughout most of the ancient world. Mortar or concrete made with lime had the disadvantage that hardening of the lime depended on the slow process of its carbonation in contact with the atmosphere. In the interior of a large mass it could take hundreds of years to harden and it would not harden under water. Sometime in the second century BC the Romans mined a pink volcanic ash from near Pozzuoli, believing it to be a sand suitable to mix with lime to produce a lime-concrete. In fact the fine volcanic ash contained silica and alumina which combined chemically with the lime to produce *pozzuolanic* cement, which set quickly independent of the atmosphere, and was stronger, more durable and had better waterproofing qualities than lime mortar. The efficacy of this hydraulic cement is well described by Vitruvius:

> There is also a kind of powder which from natural causes produces astonishing results. It is found in the neighbourhood of Baiae and in the country belonging to the towns round about Mount Vesuvius. This substance, when mixed with lime and rubble, not only lends strength to buildings, but even when piers of it are constructed in the sea, they set hard under water.

Most buildings or remains of buildings surviving from ancient times are of stone, giving, perhaps, an unbalanced impression of the amount of use made of this as a building material compared to other less durable materials, particularly timber. Nevertheless, stone found widespread use in the ancient world in the construction of structures as diverse as tombs and temples, defensive walls and pyramids, and palaces and bridges. Although availability, in the form of local geological rock strata, usually determined the type of stone used, in some cases blocks were transported over huge distances, as witness the huge granite blocks, brought from Aswan, which make up the tomb chambers of the Great Pyramid and the Roman use of marble from many countries fringing the Mediterranean.

Rock strata, which provided the stone for building, fall into three basic categories: igneous, sedimentary and metamorphic. Igneous rocks derive directly from the molten lava underlying the earth's crust, through which it has pushed up and solidified below the surface or, through volcanic activity, flowed out onto the surface. Lava not reaching the surface (intrusive) cools slowly, allowing individual crystals to form and producing a coarse-grained rock such as granite, used widely in ancient structures. Lava flowing onto the surface (extrusive) cools rapidly, producing a fine-grained rock such as basalt. The two most common sedimentary rocks used for building in the ancient world were sandstone and limestone. Sandstones are formed from sand grains (usually durable quartz grains derived from weathering of igneous rocks), bound together by cementing materials such as siliceous matter, calcium carbonate or a ferruginous compound. Limestones are formed from the precipitation in seawater of fine calcareous grains or by the accumulation on the seabed of shells and skeletons of marine organisms. Volcanic activity can also produce a sedimentary type of rock known as tuff, built up by the accumulation of falling ash, which the Romans exploited as a building material. Igneous and sedimentary rocks can be altered by great heat and pressure to form metamorphic rocks, one of which, marble, formed by the alteration of limestone, found widespread use as a building material in the ancient world.

Of the rock types used extensively in the ancient world to provide stone for building purposes, limestone was the easiest to quarry and work, first because its relative softness

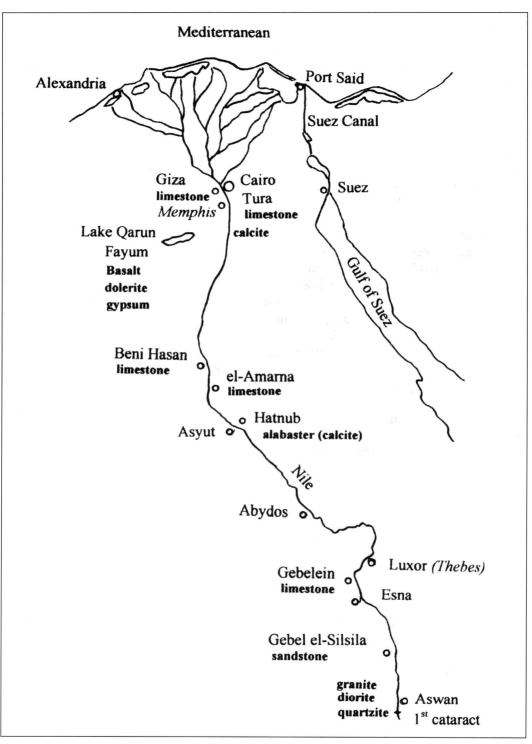

Locations of rock types used for building construction in ancient Egypt.

220

made it possible to excavate and readily trim it with the available copper or bronze tools and second because, as a sedimentary rock, it consisted of layers of sound rock interfaced with horizontal weaker bedding planes from which, after exposing the block by trenching to its full depth, blocks could be easily prised. Without the ubiquitous presence of this rock in Egypt the huge stone pyramids could not have been constructed, or certainly not on such a massive scale. The Egyptians had almost an embarrassment of excellent rock types for building stones, including, in the south of the country, sandstones used, for example, in the construction of the great New Kingdom temple at Karnak and granites used in pyramid tomb chambers and for the New Kingdom obelisks. The existence of a partly excavated, and abandoned, huge granite obelisk near Aswan has provided some indications of the methods employed in quarrying and working this rock, achieved mainly by pounding the rock with balls of diorite, an extremely hard intrusive igneous rock. The Greeks and Romans both made extensive use of marble in their building works, the Greeks using marble from Mount Pentilikon in Attica in constructing the Parthenon, and the Romans obtaining this stone from many sources bordering the Mediterranean, but mostly using it as a facing material rather than as a structural material. The Romans used copper saws to cut marble, feeding sand into the cut to act as an abrasive which embedded itself into the soft metal and is a technique used in drills and cutting discs today. Waterpower eventually replaced human effort in operating these saws, the poet Decimus Magnus Ausonius (AD 310–396) giving a graphic description of the cutting of stone slabs for the Imperial City of Treves:

> The Swift Celbis and Erubis [rivers], famed for marble, hasten full eagerly to approach with their attendant waters; renowned is Celbis for glorious fish, and that other, as he turns his millstones in furious revolutions and drives the shrieking saws through smooth blocks of marble, hears from either bank a ceaseless din.

Stone has a high compressive strength, making it very suitable for columns, but a low tensile strength, limiting its use and particularly its maximum spans when used for beams or lintels. Any flaws or incipient cracks in such beams will reduce the already low tensile strength. Deepening a beam will, in theory, increase its strength, but this advantage may be cancelled out by its own increased weight and the greater possibility of cracks or other flaws occurring in the section. The formation of a vertical crack through the depth of a deep lintel, while unsightly, may not lead to its collapse as it may act essentially as an arch, but this imposes very high horizontal compressive stresses at the top of the crack, balanced by high thrusts against adjoining lintels. The limiting span for a limestone lintel is about 3m, but a lintel of good-quality sandstone, such as that obtained from the Silsila quarries in Egypt and used in the great temple to Amon-Ra at Karnak, can span twice this distance. The granite slabs roofing the King's Tomb Chamber in the Great Pyramid span 5.2m.

While the inherent weakness and unreliability of stone beams and slabs to span distances or for roofing structures did not deter all ancient builders, other more suitable techniques based on the arch form were developed and widely used. This form of construction, if properly proportioned, ensures that the stone is always in compression. Arches, vaults and domes were usually constructed of bricks or stone blocks, in some cases

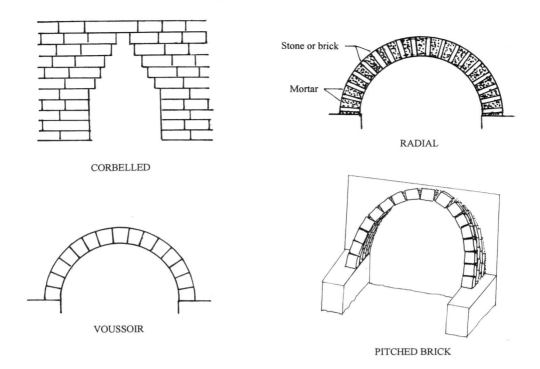

All four types of arches or vaults shown here were found in the ancient world, using either brick or stone.

achieving huge dimensions. Notable exceptions were the concrete dome of the Pantheon in Rome, and the monolithic 11m-diameter dome of the church of Theodoric in Ravenna (*c.* AD 530), fashioned from a single piece of limestone. The four different techniques of arch or vault construction listed below, using mud-brick, fired brick, or stone blocks or slabs were known, three of these dating back to at least 3000 BC and the fourth, the sophisticated *voussoir* arch, dating back to at least 700 BC.

The **corbelled**, or false arch, built up by laying bricks or stone blocks or slabs with their longest dimensions horizontally, and each layer projecting a little further into the space to be spanned than the layer immediately below it. Although this was ostensibly a cantilever construction, friction or bonding between the layers actually resulted in an action more akin to that of an arch. It had the advantage of not requiring centring or support during construction.

The **radial** arch consisting of flat bricks or pieces of stone set radially in mortar. At springing level, usually the top of a wall, the bricks were laid face down on mud mortar and canted slightly inwards by placing small stones or potsherds along the edges of the wall. By repeating this process, with each layer of bricks canted to a suitably steeper angle than the layer below, the arch was completed with the placing of the top, vertical, or near vertical, bricks. Unlike the *voussoir* arch, the mortar, not the bricks or stones, provided the truncated wedge shapes needed to achieve the curved form. In vaulted construction greater

The corbelled arch of the east gallery was built inside the 'mighty walls' of Tiryns.

Brick radial arch.

strength could be achieved by staggering the bricks in the direction of the wall, relative to the bricks in the layers immediately above and below. This type of arch or vault required support during construction in the form of timbering where available or heaped-up rubble smoothed to the arch shape at the top with a layer of mud and/or leaves, reeds or similar.

Mud-brick radial arches and vaults appear to have evolved at much the same time – around 300 BC – in Egypt and the Mediterranean. An early Egyptian example dating from about this time has been found in a tomb at Helwan. An arched gateway, giving access to the mastaba of the nobleman Neferi, who died around 2600 BC, featured bricks shaped to resemble reed bundles, clearly reflecting an earlier building technique. Early Mesopotamian examples,

In creating vaulted structures in stone, the Egyptians favoured the corbelled or false arch, as in many tombs and notably in the Grand Gallery of the Great Pyramid. However, they were aware of other arch forms, and the vaulting for the storerooms associated with the Ramesseum temple of Rameses II is a spectacular example, using mud-brick, of the pitched brick technique.

224

besides those at Ur, include arches and vaults with spans of up to 3.8m in a temple complex found at Tell al Rimah dating from about 2000 BC. Mud-brick radial arches found in Israel, dating variously between the eighteenth and thirteenth centuries BC, include one arch ring consisting of three concentric courses spanning an impressive 4.25m.

The **pitched brick** technique, used mostly for vaulting, consisted of rings of thin bricks laid on edge with their smallest dimensions in the longitudinal direction of the vault, the bricks canted back slightly to rest on the ring previously constructed. In the case of the Arch of Ctesiphon the bricks, about 300mm square and 75mm thick, are canted back at an angle of about 18° from the vertical. With mortar prepared to the right consistency, the newly placed bricks would be held to the bricks of the previous ring by moisture suction in the mortar, until it dried, giving a strong joint. This ingenious method did not require centring. The vault started by matching it to a marked outline on an end wall, building up successive partial rings from the base, each slightly canted back, until a full ring was completed with its crest touching the wall and its springing some distance out from the wall. Vaulting then proceeded by constructing similar rings, each resting on the previous one, until, finally, a procedure analogous to that at the start completed the vault to a vertical face. Usually the vault thickness was made up of a number of layers of bricks.

The *voussoir* arch consisted of truncated wedge-shaped bricks or stone blocks. With carefully shaped *voussoirs* this type of arch or vault did not need mortar, but it did need centring or support during construction.

Roman voussoir arches were not always composed of finely finished and uniform size blocks, but still remain remarkably stable, as seen in this town gate today.

By the time of Herodotus, two main Orders of Architecture had emerged, the Doric, which developed on the Greek mainland, and the Ionic in Asia Minor. Although both adopted the basic rectangular plan, a number of differences distinguished them, in addition to the frequently cited column capitals. The plainer Doric capital consisted of the inverted mushroom-shaped *echinus* bearing against the square *abacus*, on which rested the *architrave*, a plain beam or lintel which spanned the columns and provided the basic support for the roof structure; the Ionic capital featured a capital in the shape of an inverted scroll, or *volute*, bearing against a much thinner *abacus* on which rested a horizontally banded *architrave*. At their bases the Doric columns rested directly on the slabbed flooring or *stylobate*, whereas each Ionic column rested on a round base bearing on the *stylobate*. The columns in both cases consisted of fluted drums; but whereas the grooves forming the flutes met in sharp lines in the Doric columns, narrow flat strips in the Ionic columns separated them. Ionic columns were generally higher and more slender than their bulky Doric counterparts, which rarely had a ratio of diameter to height of less than one to six, reflecting an increased confidence on the part of the builders in the ability of stone to support heavy loads.

Immediately above the *architrave* in each order was a frieze surmounted by a cornice, the latter forming the eaves along the length of the building and the bases of the gables at the ends of the building. The friezes in the Doric, consisting of alternating short, vertically grooved (*trigliths*) and plain (*metopes*) lengths, contrasted with the (usually) wholly plain

Doric   Ionic   Corinthian

Classic orders of architecture.

friezes of the Ionic. In special circumstances relief sculptures were used to decorate the *metopes* or the plain Ionic friezes. Double sloping courses known as 'raking cornices' and 'raking *simas*' defined the upper edges of the triangular end gables, or *pediments*. The roof itself, usually of overlapping terracotta pan tiles with upturned edges supporting cover tiles, rested on a timber framework.

Although the Doric continued strongly in use through the Hellenistic period (*c.* fourth to first century BC), it then lost its appeal, in contrast to the Ionic and its variant, the even more slender Corinthian, which continued in use throughout the Roman Empire. The more elaborate leafy capitals of the Corinthian evolved in stages from the Ionic, notwithstanding the rather charming alternative explanation offered by Vitruvius:

A freeborn maiden of Corinth, just of marriageable age, was attacked by an illness and passed away. After her burial, her nurse, collecting a few little things that gave the girl pleasure while she was alive, put them in a basket, carried it to the tomb, and laid it on top thereof, covering it with a roof-tile so that the things might last longer in the open air. This basket happened to be placed just above the root of an acanthus. When springtime came round, the root of the acanthus, being pressed down in the middle by the weight, produced leaves and shoots which grew up the sides of the basket and, being pressed down at the angles by the weight of the tile, the stalks were compelled to bend into volutes at their ends. Then Callimachus, whom the Athenian nick-named *catatechnos* for the refinement and delicacy of his marble carving, passed by this tomb and saw the basket with the tender young leaves growing round it. Delighted by their style and novelty, he built some columns with this form for the Corinthians and fixed from that time the rules for works of the Corinthian order.

SELECT BIBLIOGRAPHY

Adam, Jean-Pierre, *Roman Building: Materials and Techniques* (trans. Anthony Mathews). London, B.T. Batsford, 1994.

Anon., *The Grand Canal of China*. South China Morning Post Ltd and New China News, 1984.

Armayor, O. Kimball, *Herodotus' Autopsy of the Fayoum: Lake Moeris and the Labyrinth of Egypt*. Amsterdam, J.C. Gieben, 1985.

Atkinson, R.J.C., *Stonehenge*. London, Hamish Hamilton, 1956.

Atkinson, Richard, *Silbury Hill*. Chronicle, London, 1956. BBC, 1978, pp. 159–73.

Baines, J. and Malek, J., *Atlas of Ancient Egypt*. Oxford, Facts on File, 1992.

Barnatt, John, *Stone Circles of the Peak*. London, Turnstone Books, 1978.

Bellonci, Maria, *The Travels of Marco Polo* (trans. Teresa Waugh). London, Book Club Associates, 1984.

Brohyier, R.L., *Ancient Irrigation Works in Ceylon*. Colombo, Sri Lanka Ministry of Mahaweli Development, 1934.

Bromwich, James, *The Roman Remains of Southern France*. London, Routledge, 1993.

Brown, David J., *Bridges*. London, Mitchell Beazley, 1993.

Burl, Aubrey, *Prehistoric Avebury*. New Haven, Yale University Press, 1979.

Bury, J.B., *A History of Greece*. London, Macmillan, 1931.

de Camp, L. Sprague, *The Ancient Engineers*. Adelaide, Rigby, 1963.

Casson, Lionel, *Travel in the Ancient World*. London, George Allen Unwin, 1974.

Ceram, C.W., *The First American*. New York, Harcourt Brace Jovanovich, 1971.

Chevallier, Raymond, *Roman Roads*. London, B.T. Batsford, 1976.

Cook, J.M., *The Persians*. London, Folio Society, 1982.

Cotterell, Brian and Kamminga, Johan, *Mechanics of Pre-Industrial Technology*. Cambridge, Cambridge University Press, 1990.

Coulton, J.J., *Ancient Greek Architects at Work*. New York, Cornell University Press, 1977.

Dames, Michael, *The Silbury Treasure*. London, Thames and Hudson, 1976.

Davey, Norman, *A History of Building Materials*. London, Phoenix House, 1961.

Doe, Brian, *Monuments of South Arabia*. Cambridge, Falcon-Oleander Press, 1983.

Edwards, I.E.S., *The Pyramids of Egypt*. Harmondsworth, Penguin, 1961.

Evans, J.D., *Malta*. London, Thames and Hudson, 1959.

Farrington, I.S., 'The Archaeology of Irrigation Canals', *World Archaeology/Water Management*, 2/3 (1980), 287–305.

Forbes, R.J., *Studies in Ancient Technology*, vols I & II. Leiden, E.J. Brill, 1964–5.

Franck, Irene M. and Brownstone, David M., *The Silk Road: A History*. New York, Facts on File, 1986.

Fugal-Mayer, H., *Chinese Bridges*, Shanghai, Kelly and Walsh, 1937.

Garbrecht, Günther, 'Hydrologic and Hydraulic Concepts', in *Antiquity, Hydraulics and Hydraulic Research*. Rotterdam, A.A. Balkema, 1987, pp. 1–22.

Garlake, Peter S., *Great Zimbabwe*. London, Thames and Hudson, 1973.

Gates, Charles, *Ancient Cities*. London, Routledge, 2003.

Goodfield, Jane, 'Tunnel of Eupalinus', *Scientific American*, 210/6 (June 1964).

Hagen, Victor, W. von, *The Roads that Led to Rome*. London, Weidenfeld and Nicolson, 1967.

Hamblin, Dora Jane, *The First Cities*. New York, Time-Life Books, 1973.

Hammerton, J.A., *Wonders of the Past* (three volumes). London, Fleetway House, 1933.

Hayes, W.C., *The Sceptre of Egypt*, pts I and II, New York, Metropolitan Museum of Art, 1953, 1959.

Hindley, Geoffrey, *A History of Roads*. London, Peter Davies, 1971.

Hoare, R.C., *The Ancient History of South Wiltshire*. London, 1812.

Hodge, A. Trevor, *Roman Aqueducts and Water Supply*. London, Duckworth, 1992.

Holroyd, Stuart and Lambert, David, *Mysteries of the Past*. London, Aldus Books, 1979.

Humphrey, J.W., Oleson, J.P. and Sherwood, A.N., *Greek and Roman Technology: A Source Book*. London, Routledge, 1998.

Hyslop, John, *The Inca Road System*. New York, Academic Press, 1984.

Johnson, Stephen, *Hadrian's Wall*. London, B.T. Batsford/English Heritage, 1989.

Kenyon, Kathleen, *Digging up Jericho*. London, Ernest Benn, 1957.

Kienast, Hermann J., *Die Wasserleitung des Eupalinos auf Samos*. Athens, Kulturministerium Kasse Archäologische Mittel und Entiegnungen, 2004.

Kirby, R.S., Withington, S., Darling, A.R. and Kilgour, F.G., *Engineering in History*. New York, McGraw Hill, 1956.

Koldewey, Robert, *The Excavations at Babylon* (trans. Agnes S. Johns). London, Macmillan, 1914.

Lepper, Frank and Frere, Sheppard, *Trajan's Column*. Stroud, Alan Sutton, 1988.

Lloyd, S., Müller, H.W. and Martin, R., *Ancient Architecture: Mesopotamia, Egypt, Crete, Greece*. New York, Harry N. Abrams, 1974.

Luo, Zhewen and Zhao Luo, *The Great Wall of China in History and Legend*. Beijing, Foreign Languages Press, 1986.

Malone, Caroline, *Avebury*. London, B.T. Batsford/English Heritage, 1989.

Mao, Yi-sheng, *Bridges in China: Old and New*. Peking, Foreign Languages Press, 1978.

Margary, Ivan, D., *Roman Roads in Britain*. London, Phoenix House, 1955.

Marlowe, John, *The Making of the Suez Canal*. London, Cresset Press, 1964.

Mee, Christopher and Spawforth, Antony, *Greece: An Oxford Archaeological Guide*. Oxford, Oxford University Press, 2001.

Mendelssohn, K., *The Riddle of the Pyramids*. London, Thames and Hudson, 1974.

Millward, Roy and Robinson, Adrian, *The Peak District*. London, Eyre Methuen, 1975.

Moseley, Michael E., *The Incas and their Ancestors*. London, Thames and Hudson, 1992.

Murray, M.A., *The Splendour that was Egypt*. London, Sidgwick and Jackson, 1964.

Myres, J.L., *Herodotus: Father of History*. Oxford, Clarendon Press, 1953.

Needham, Joseph, Lu Gwei-Djen and Ling Wang, *Science and Civilisation in China*, vol. IV, no. 3. Cambridge, Cambridge University Press, 1971.

Neuburger, Albert, *The Technical Arts and Sciences of the Ancients* (trans. Henry L. Brose). London, Methuen, 1930.

Nuttgens, Patrick, *The Story of Architecture*. London, Phaidon, 1993.

Oates, Joan, *Babylon*. London, Thames and Hudson, 2000.

O'Kelly, Michael, J., *Newgrange*. London, Thames and Hudson, 1982.

Park, Chris, 'Water Resources and Irrigation Agriculture in Pre-Hispanic Peru', *Geographical Journal* , 149/2 (July 1983), 153–66.

Parry, Dick, 'Megalith Mechanics', *Civil Engineering*, 138/4 (2000), 183–92.

Parry, Dick, *Engineering the Pyramids*. Stroud, Sutton, 2004.

Pavel, P., 'Raising the Stonehenge Lintels in Czechoslovakia', *Antiquity*, 66/251 (1992), 389–483.

Payne, Robert, *The Canal Builders*. New York, Macmillan, 1959.

Petrie, W.M.F., *The Pyramids and Temples of Gizeh*. London, 1883.

Pitts, Mike, *Hengeworld*. London, Century, 2000.

Richards, Julian, *Stonehenge*. London, B.T. Batsford/English Heritage, 1991.

Richards, J. and Whitby, M. 'The Engineering of Stonehenge', in B. Cunliffe and C. Renfrew (eds), *Science and Stonehenge*, Oxford, Oxford University Press, for the British Academy, 1997.

Sandström, Gösta, E., *The History of Tunnelling*. London, Barrie and Rockliff, 1963.

Schnitter, Nicholas, J., *A History of Dams*. Rotterdam, A.A. Balkema, 1994.

Schoder, S.J., Raymond, V., *Ancient Greece from the Air*. London, Thames and Hudson, 1974.

Schreiber, Hermann, *The History of Roads* (trans. Stewart Thomson). London, Barrie and Rockliff, 1961.

de Sélincourt, Aubrey, *The World of Herodotus*. London, Secker and Warburg, 1962.

Sellin, Robert H.J., 'The Large Roman Water Mill at Barbegal', *Houille Blanche*, No. 6, 1981.

Sherratt, Andrew (ed.), *The Cambridge Encyclopedia of Archaeology*. Cambridge, Cambridge University Press, 1980.

Simpson, J.H., 'Further Reflections on Megalith Mechanics', *Civil Engineering*, 144/4 (November 2001), 181–5.

Singer, Charles, Holmyard, E.J., Hall, A.R. and Williams, Trevor I., *A History of Technology* (vols I & II). Oxford, Clarendon Press, 1954.

Sitwell, N.H.H., *Roman Roads of Europe*. London, Cassell, 1981.

Smith, Isobel F., *Windmill Hill and Avebury*. London, Clover Press, 1959.

Smith, H. Shirley, *The World's Great Bridges*. London, Phoenix House, 1964.

Sowers, G.F., 'There were Giants in those Days', *ASCE Journal of the Geotechnical Division*, 107/GT4 (1981), 385–419.

Stone, E. Herbert, *Stonehenge*. London, Robert Scott, 1924.

Strandh, S., *A History of the Machine*. New York, A & W Publishers, 1979.

Straub, Hans, *A History of Civil Engineering* (trans. E. Rockwell). London, Leonard Hill, 1960.

Taylour, Lord William, *The Mycenaens*. London, Thames and Hudson, 1964.

Toy, Sidney, *A History of Fortification*. London, William Heinemann, 1955.

Trombold, Charles D., *Ancient Road Networks and Settlement Hierarchies in the New World*. Cambridge, Cambridge University Press, 1991.

Trump, D. H., *Malta: An Archaeological Guide*. London, Faber and Faber, 1972.

Ubbelohde-Doering, H., *On the Royal Highways of the Inca* (trans. Margaret Brown). London, Thames and Hudson, 1967.

Van Slyke, and Lyman. P., *Yangtze*. Stanford, Addison-Wesley, 1988.

White, K. D., *Greek and Roman Technology*. London, Thames and Hudson, 1986.

Whittle, Alasdair, *Sacred Mounds, Holy Rings*. Monograph 74, Oxford, Oxbow Books, 1997.

Wilson, Sir Arnold, T., *The Suez Canal*. London, Oxford University Press, 1939.

Winslow, E. M., *A Libation to the Gods*. London, Hodder and Stoughton, 1963.

Woolley, Sir Leonard, *Excavations at Ur*. London, Ernest Benn, 1954.

Woolley, Sir Leonard, *History Unearthed*. London, Ernest Benn, 1958.

Zewen, Luo, Wilson, Dick, Drege, J. P. and Delahaye, H., *The Great Wall*. London, Michael Joseph, 1981.

CLASSICAL WRITERS

Ausonius, *Moselle* (trans. H.G. Evelyn White). Loeb Classical Library, Harvard University Press.

Caesar, *The Gallic War* (trans. H.J. Edwards). Loeb Classical Library, Harvard University Press.

Dio Cassius, *Roman History* (trans. E. Cary). Loeb Classical Library, Harvard University Press.

Diodorus Siculus, *Books I & II* (trans. C.H. Oldfather). Loeb Classical Library, Harvard University Press.

Frontinus, *Stratagems and Aqueducts* (trans. Charles E. Bennett). Loeb Classical Library, Harvard University Press/Heinemann.

Herodotus, *The Histories* (trans. Aubrey de Sélincourt). Harmondsworth, Penguin Books (rev. edn by A.R. Burn).

Horace, *Satires, Epistles, Ars Poetica* (trans. H.R. Fairclough). Loeb Classical Library, Harvard University Press/Heinemann.

Josephus, *Jewish Antiquities*. Loeb Classical Library, Harvard University Press.

Pliny the Elder, *Natural History* (*A Selection*, trans. John F. Healy). London, Penguin Books.

Statius, *Silvae: Thebaid I–IV* (trans. J.H. Mozley, 1982). Loeb Classical Library, Harvard University Press/Heinemann.

Strabo, *Geography Books 15–17* (trans. Horace Leonard Jones). Loeb Classical Library, Harvard University Press.

Suetonius, *The Twelve Caesars*. London, Penguin Books.

Tacitus, *Annals* (trans. J. Jackson). Loeb Classical Library, Harvard University Press.

Vitruvius, *The Ten Books of Architecture* (trans. Morris Hickey Morgan). London, Harvard University Press.

Xenophon, *The Persian Expedition* (trans. Rex Warner). London, Penguin Books.

Sources of Quoted Passages

Antipater of Thessolonika, *Greek Anthology*, 9.418 (p. 9)

Ausonius, *Moselle*, 359–64 (p. 221)

Dio Cassius, *Roman History*, 71.3.1 (p. 67)

Diodorus Siculus, *History*, 1.10.2–6 (p. 186)

Herodotus, *The Histories*, 1.93–4 (pp. 210–11); 1.98 (p. 113); 1.178–81 (p. 106); 1.181–2 (p. 130); 1.191 (p. 4); 1.193 (p. 19); 1.194 (p. 39); 2.99 (pp. 1–2); 2. 124–5 (pp. 201–2); 2.137–8 (pp. 162); 2.148 (p. 179); 2.175 (p. 162); 3.60 (p. 34); 5.52 (p. 81); 7.22–4 (pp. 52–3); 7.33–6 (p. 64)

Horace, *Satires*, 1.5 (p. 42)

J. Needham, Lu Gwei-Djen and Ling Wang, *Science and Civilisation in China*, 4.3.51–2 (p. xi); 4.3.178 (p. 78); 4.3.295 (p. 28); 351 (p. 46)

Statius, *Silvae*, 4.3.40–55 (p. 94)

Strabo, *Geography*, 16.1.9 (p. 130)

Tacitus, *Annales*, 13.53 (p. 57)

Vitruvius, *The Ten Books on Architecture*, 2.6.1 (p. 219); 2.8.18–19 (p. 218); 2.9.6–8 (p. 214); 4.1.9–10 (p. 227); 10.2.11–12 (p. 167)

INDEX OF NAMES

INDEX OF PLACES

GENERAL INDEX